MASTERY OF JOINERY AND BUSINESS

The Modern Carpenter Joiner
and Cabinet-Maker

MASTERY OF JOINERY AND BUSINESS

G. Lister Sutcliffe, Editor
Associate of the Royal Institute of British Architects,
member of the Sanitary Institute, editor and joint-author of
Modern House Construction, author of *Concrete:*
Its Nature and Uses

Roy Underhill, Consultant
Television host of "The Woodwright's Shop",
author of *The Woodwright's Shop*, *The Woodwright's*
Companion, and *The Woodwright's Workbook*, and
Master Housewright at Colonial Williamsburg

A Publication of
THE NATIONAL HISTORICAL SOCIETY

The Modern Carpenter Joiner and Cabinet-Maker presented
up-to-date techniques and tools for its time. However, much has
changed since 1902. Not all materials and methods described in these
pages are suitable for the construction materials and tools of today.
Before undertaking any of the building, remodeling or other practices
described in these pages, the reader should consult with a reputable
professional contractor or builder, especially in cases where
structural materials may come under stress and where structural
failure could result in personal injury or property damage. The
National Historical Society, Cowles Magazines, Inc., and Cowles Media
Company accept no liability or responsibility for any injury or loss
that might result from the use of methods or materials as described
herein, or from the reader's failure to obtain expert professional
advice.

Library of Congress Cataloging-in-Publication Data
Mastery of joinery & business / G. Lister Sutcliffe,
    editor; Roy Underhill, consultant.
        p. cm. — (The Modern carpenter joiner and cabinet-maker)
    ISBN 0-918678-62-5
        1. Joinery. 2. Construction industry — Management. 3. Build-
ing-Estimates. 4. Construction contracts.
I. Sutcliffe, G. Lister. II. Underhill, Roy. III. Title: Mastery of joinery
and business. IV. Series.
    TH5662.M37   1990
    694'.6 — dc20   90-6303
                                                                      CIP

CONTENTS

DIVISIONAL-VOL. VIII

ILLUSTRATIONS

DIVISIONAL-VOL. VIII

ILLUSTRATIONS IN TEXT

SECTION XIII.—SHOP MANAGEMENT

SECTION XIV.—ESTIMATING

GLOSSARY

The Glossary is illustrated by a series of 118 wood-cuts elucidating the terms that require such aid to make them fully understood.

PREFACE

"There! I've finished my door to-day, anyhow," called out the distracted young joiner in the opening pages of George Eliot's 1896 novel *Adam Bede*. To his embarrassment, and to his co-workers' amusement, he was holding up for approval an empty frame, having forgotten to make and fit the panels. This final volume of *The Modern Carpenter Joiner and Cabinet-Maker* devotes much of its ink to door-making and panel-frame construction. Here, Sutcliffe observes the imperfection of almost every method of joining timber, owing to the tendency of wood to shrink or swell in response to dry or damp air. In most work, this instability is of small consequence, but in a door it can be totally defeating. Winter dryness can make a door shrink and open up cracks that let in the cold; summer humidity can make it swell and stick so tight that it cannot be opened. By far, Sutcliffe reminds us, "the most perfect mode of satisfying both conditions is by the use of framed work." But, like the joiner who forgot the panels, we forget how the design of the familiar panel-frame door evolved directly from the quest to solve the problems of water and wood.

The idea for framed work is ancient. Frame carpentry began with simple skeletal structures of timber, the spaces between filled with interwoven sticks plastered with clay. When this design is extended to joinery, however, it perfectly solves the problem of moving wood. Although wood shrinks and swells across the grain with changes in humidity, it is remarkably stable along its length. By defining the perimeter of a door or wainscot with the long grain of the wood, it will not swell or shrink in its outer dimensions. The broad panels set in grooves cut for them within this stable outer frame can then expand and contract harmlessly. The result is a door that does its job, never sticking or cracking with the weather.

Of course, for this strategy to work, the wood must be free to move. Workers who forget the fundamental principle of panel-frame construction doom their doors to destruction by overuse of the glue pot. The 1734 *Builder's Dictionary* warned joiners to leave their panels free of glue and nails, because: "This will give Liberty to the Board to shrink, and swell without tearing; wheras Mouldings that are nailed round the Edge, as the common Way is, do so restrain the Motion of the Wood, that it cannot shrink without tearing."

Although wood will always respond to moisture changes in the surrounding air, it must still be as dry as possible before use. Wood is always responsive, but, as one builder put it, "this evil will be enhanced in proportion as the wood is less seasoned." Shakespeare showed his awareness of the problem when he compared joinery with unseasoned timber to a mock wedding in *As You Like It;* "this fellow will but join you together as they join wainscote; then one of you shall prove a shrunk panel and, like green timber, warp, warp."

The seasoning of the material is so important that it is common practice to make doors in stages that allow for "second seasoning." The joiner brings the material almost to the final cut, and then sets it aside for about a week before continuing. Oak is particularly prone to moving after a fresh surface is exposed. The *Builder's Dictionary* warned of this: "For it has been observ'd, that tough Boards have lain in an House ever so long, and are ever so dry, yet when they are thus shot and planed, they will shrink afterwards beyond Belief." The warnings in this series against using unseasoned stock are ancient indeed.

In the days of the village joiner's shop, making a basic, four-panel door was considered a good day's work. This meant ripping, planing, and grooving the stiles and rails of the frame, laying out and cutting the ten mortice-and-tenon joints, planing and molding the four panels to fit into the grooves of the frame, and then fitting the whole together. This job admits not the slightest carelessness. As Nicholson's 1860 *Dictionary of Architecture* cautioned, both the tools and the workman must be sharp and true, "these being strictly attended to, the work will of necessity, when put together, close with certainty; but if otherwise, the workman must expect a great deal of trouble." Although staircase and handrailing work may be more mathematically demanding, a well-made door is most worthy of respect; truest to the nature of wood, and a fitting end to this series.

ROY UNDERHILL
MASTER HOUSEWRIGHT
COLONIAL WILLIAMSBURG

Section XII—JOINERY

BY

THE EDITOR

Section XII—JOINERY

CHAPTER I

MOULDINGS

Before entering on the consideration of the practical part of the subject of joinery, it may be well to illustrate and describe the various ornamental mouldings which may have to be formed by the joiner.

When any moulding is formed on a piece of framing it is said to be *on the solid* or *stuck*. When it is formed on a separate piece of wood, and attached to the part of the framing which it is meant to ornament, it is said to be *laid in* or *planted*.

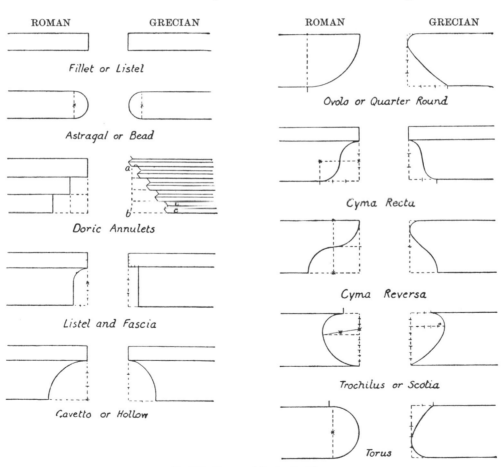

Fig. 1081.—Roman and Grecian Mouldings

Fig. 1081 contains some Grecian and Roman versions of the same mouldings. Other examples will be found in Plates I, II, and III, and figs. 7 and 13, Section I.

Fillet or Listel is a right-angled member, and requires no description.

The Astragal or Bead.—To describe this moulding, divide its height into two equal parts, and from the point of division as a centre describe a semicircle, which is the contour of the astragal.   The *bead* is of constant occurrence in joinery.   When the edge of a

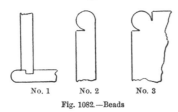

No. 1   No. 2   No. 3

Fig. 1082.—Beads

piece of wood is reduced to a semi-cylindrical form, as in fig. 1082, No. 1, it is said to be rounded.   When the rounding forms more than a semicircle, and there is a sinking on the face, as in No. 2, the rounding is termed a *quirk-bead*, the groove or sinking being termed the *quirk*.   When the edge is rounded with a sinking or groove on both faces, as in No. 3, the moulding is a double-quirk bead.

Doric Annulets.—The left-hand figure shows the Roman, and the right-hand figure the Grecian form of this moulding.   To describe the latter, divide the height *b a* into four equal parts, and make the projection equal to three of them.   The vertical divisions give the lines of the under side of the annulets, and the height of each annulet, *c c*, is equal to one-fifth of the projection; the upper surface of *c* is at right angles to the line of slope.

Listel and Fascia.—Roman: Divide the whole height into seven equal parts, make the listel equal to two of these, and its projection equal to two.   With the third vertical division as a centre, describe a quadrant.   Grecian: Divide the height into four equal parts, make the fillet equal to one of them, and its projection equal to three-fourths of its height.

Cavetto or Hollow.—In Roman architecture this moulding is a circular quadrant; in Grecian architecture it is an elliptical quadrant, which may be described by any of the methods given in Section V.

Ovolo or Quarter-round.—This is a convex moulding, the reverse of the cavetto, but described in the same manner.   The Grecian forms of this and of the torus are often known as "thumb" mouldings.

Cyma Recta.—A curve of double curvature, formed of two equal quadrants.   In the Roman moulding these are circular, and in the Grecian moulding elliptical.

Cyma Reversa.—A curve of double curvature like the former, but reversed, and formed in the same manner.

Trochilus or Scotia.—A hollow moulding, which in Roman architecture may be formed of two unequal circular arcs, thus:—Divide the height into ten equal parts, and at the sixth division draw a horizontal line.   From the seventh division as a centre, and with seven divisions as radius, describe from the lower part of the moulding an arc, cutting the above horizontal line, and join the centre and the point of intersection; bisect this line, and from the point of bisection as a centre, with half the length of the line as radius, describe an arc to form the upper part of the curve.   There are many other methods of drawing this moulding.   The Grecian trochilus is an elliptical or parabolic curve; the proportions of one form of it are shown by the divisions of the dotted lines.

The Torus.—The Roman moulding is semi-cylindrical, and its contour is, of course, a semicircle.   The Grecian moulding is either elliptical or parabolic; and although this and the other Greek mouldings may be drawn, as we have said, by one or other of the methods of drawing ellipses and parabolas, and by other methods about to be illustrated, it is much better to become accustomed to sketch them by the eye, first setting off their projections, as shown in fig. 1081, by the divisions of the dotted lines.

Fig. 1083—*The Quirked Ovolo.*—The projection of the moulding in No. 1 is made equal to five-sevenths of its height, as seen by the divisions, and the radius of the circle *b c* is made equal to two of the divisions, but any other proportions may be taken.   Describe the circle *b c*, forming the upper part of the contour, and from the point *g* draw *g h* to form a tangent to the lower part of the curve.   Draw *g a* perpendicular to *g h*, and make *g f* equal to the radius *d c* of the circle *b c*, join *f d* by a straight line, which bisect by a line perpendicular to

it, meeting *g a* in *a*.   Join *a d*, and produce the line to *c*.   Then from *a* as a centre, with the radius *a c* or *a g*, describe the curve *c y*.

No. 2.—*To draw an ovolo, the tangent* d e *and point of extreme projection* b *being given.*

Through *b* draw the vertical line *g h*, and through *b* draw *b c* parallel to the tangent *d e*, and draw *d c* parallel to *g h*, and produce it to *a*, making *c a* equal to *c d*.   Divide *e b* and *c b* each into the same number of equal parts, and through the points of division in *c b* draw from *a* right lines, and through the points of division in *e b* draw from *d* right lines, cutting those drawn from *a*.   The intersections will be points in the curve.   The proportions

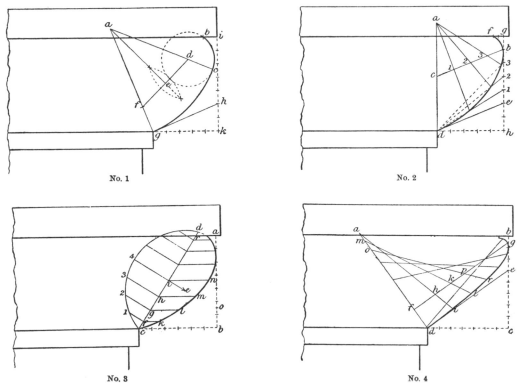

Fig. 1083.—Four Methods of drawing the Grecian Ovolo

may be varied by altering the projection and the angle which the tangent makes with the horizontal line *d h*.

No. 3.—Divide the height *b a* into seven equal parts, and make *a r* equal to one and a half of these parts, as *b o*, and make *b c* equal to six parts; join *c r*, and produce it to *d*, and make *c d* equal to eight and a half divisions.   Bisect *c d* in *i*, and draw through *i*, 4 *i* at right angles to *c d*, and produce it to *e*; make *i e* equal to *b o*, and from *e* as a centre, with radius *e c* or *e d*, describe the arc *c d*.   Then divide the arc into equal parts, and draw ordinates to *c d*, in 1 *f*, 2 *g*, 3 *h*, 4 *i*, &c., and corresponding ordinates *f k* (equal to 1 *f*), *g l* (equal to 2 *g*), &c., to find points in the curve.

No. 4.—The height is divided into eight equal parts, seven of which are given to the projection *d c*.   Join *d* and the fifth division *e*, and draw *d a* at right angles to *d e*.   Make *d f* equal to two divisions, and draw *f g* parallel to *d e*, then *d f* is the semi-axis minor, and *f g* the semi-axis major of the ellipse; and the curve can either be trammelled or drawn by means of the lines *a h*, *m k*, *o p*, being made equal to the difference between the two semi-axes, and being produced to intersect the arcs drawn from *a*, *m*, *o* as centres with radius *f g*.

The same methods may be adopted for drawing the Grecian torus; this will be obvious if fig. 1083 is turned upside down.

Fig. 1084.—*To describe the hyperbolic ovolo of the Grecian Doric capital, the tangent* d f *and projection* h *being given.*

Draw *d e g k a* perpendicular to the horizon, and draw *g h* and *e f* at right angles to *d e g k a*. Make *g a* equal to *d g*, and *g k* equal to *d e*; join *h k*. Divide *h k* and *f h* into the same number of parts, and draw lines from *a* through the divisions of *k h*, and lines from *d*

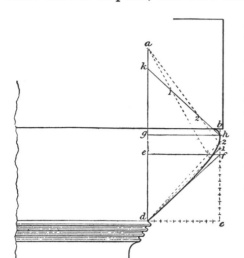

through the divisions of *f h*, and their intersections are points in the curve.

Fig. 1085, No. 1, is the Roman trochilus. Bisect the height *h b* in *e*, and draw *e f* parallel to the horizontal lines and cutting *g c* in *f*; divide the projection *h g* into three equal parts, make *e o* equal to one of the divisions, and *f d* equal to two of them, join *d o*, and produce the line to *a*. Make *d c* equal to *d g*, and draw *c b*, and produce it to *a*. Then from *d* as a centre, with radius *d a* or *d g*, describe the arc *g a*; and from *o* as a centre, with radius *o a*, describe the arc *a b*.

Fig. 1084.—The Doric Capital

No. 2 shows the method of drawing the Grecian trochilus by intersecting lines in the same manner as the rampant ellipse, fig. 393, page 254, Vol. I.

Fig. 1086, No. 1, shows the cyma recta formed by two equal opposite curves, imitations of the ellipse, drawn in the manner taught in Problem LXXX, page 256, Vol. I. By taking a greater number of points as centres, a figure resembling still more closely the true elliptical curve will be produced. No. 2 shows the cyma recta formed with true elliptical quadrants, described as in fig. 403, page 256, Vol. I.

No. 3 shows the cyma reversa, obtained in the same manner. The lines *c m*, *k h* are the semi-axes major, and the lines *k l*, *l m*, represent the two semi-axes minor. As in the former case, these lines, and the heights *k l*, *m l*, being obtained, the curve can be trammelled.

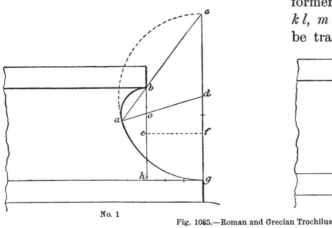

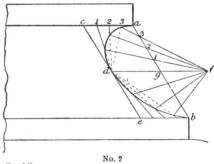

No. 1   No. 2

Fig. 1085.—Roman and Grecian Trochilus

No. 4 shows the cyma recta used as a base moulding, and obtained by means of ordinates from circular arcs, as in No. 3, fig. 1083.

It is obvious that the relation of the projection to the height in all these mouldings may be infinitely varied; but if the student has paid attention to the construction of the ellipse, parabola, and hyperbola, elucidated in pages 252–258, Vol. I, these variations will not embarrass him. But it is necessary to repeat that it is better, after the eye has become familiarized with the graceful forms of the Grecian mouldings, to trust to the curves produced by sketching, as then the proportions may be varied to infinity.

The mouldings just described are generally denominated *classic*, and are derived from examples left us by the Greeks and Romans. But the architecture of the middle ages contains mouldings which are no less attractive in their forms, and of which some typical examples have been given in Plate VIII.

PLATE LXXXIX

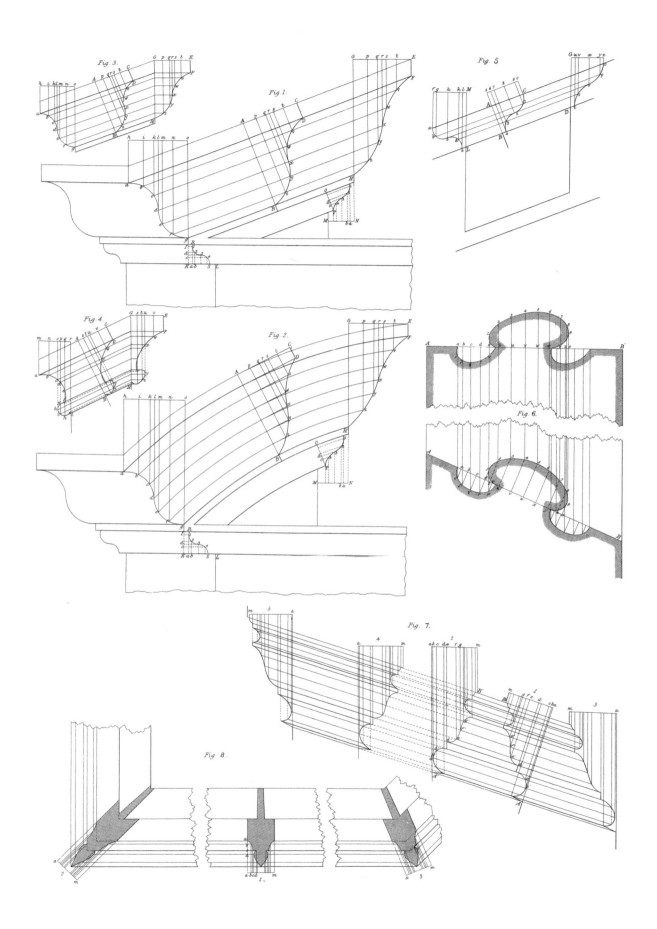

RAKING MOULDINGS

Raking Mouldings—Plate LXXXIX.—Fig. 1 shows part of the raking cornice of a pediment, with the horizontal part of the moulding on the left of the figure. Draw *g o* perpendicular to the horizon, and *o h* at right angles to *g o*. In *o h* take any point, *l*, and draw *l d* parallel to *g o* and cutting the profile in *d*, and through *d* draw a line *d x* parallel to the line of rake. Then to find the section of the raking front, draw any line, A B, perpendicular to *d x*, and make A *r* equal to *o l*, and draw *r x* parallel to A B, cutting *d x* in *x*; then the point *x* is a point in the raking profile. In the same manner any other points, such as *z, y, w, v*, corresponding to *f, e, c, b*, may be found.

When the moulding is returned at the upper part, as at H F, the line H G must obviously

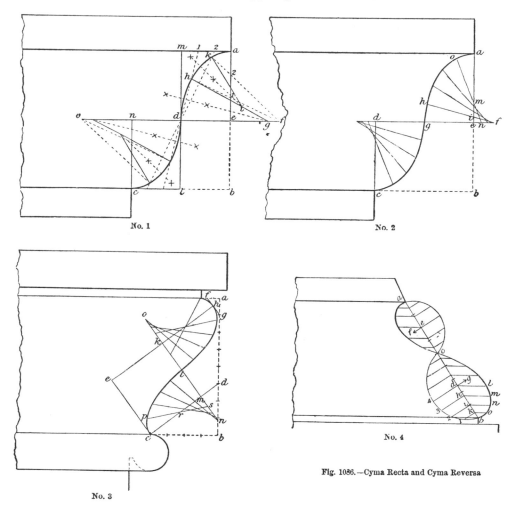

No. 1

No. 2

No. 3

No. 4

Fig. 1086.—Cyma Recta and Cyma Reversa

be drawn parallel to *g o*, that is, perpendicular to the horizon. The remainder of the procedure, and the manner of finding the return of the bed moulding R S at H P, is obvious.

Fig. 2 shows a raking moulding on the spring. In this the procedure is the same as in the last, except that in place of drawing lines parallel to the rake, concentric curves are described to find the points in the moulding. But it is necessary to observe that it is not where the perpendiculars from A C intersect these arcs that the proper points are. The true points are intersections with tangents to the curves where they cut the line A B.

Figs. 3 and 4 show the method of describing the section of the raking moulding on the line A B perpendicular to the raking line, and also on the line G H parallel to *g o*, in the case where the moulding is not returned, or where the two raking sides meet.

Fig. 5 shows the manner of drawing a modillion in a raking cornice. The mode of operating is precisely as in fig. 1. Fig. 6 is a raking architrave, and fig. 7 a base moulding; in the latter the true section of the raking moulding is shown at 1, return

mouldings forming external mitres are shown at 2 and 3, and others forming internal mitres at 4 and 5.

Fig. 8 shows the method of describing the angle-bars of a window—No. 1 being the ordinary bar; No. 2, a mitre bar; and No. 3, a bar occupying an obtuse angle. The line *a m* is in all three cases drawn perpendicular to the axis of the bar; and the divisions *a b c d*, &c., of No. 1 are transferred to Nos. 2 and 3, and lines perpendicular to *a m* drawn to meet the lines *a b c d*, which pass through points of the moulding.

Enlarging and Diminishing Mouldings—PLATE XC, fig. 1.—Let A B be the height of a cornice which it is proposed to diminish. On A B construct an isosceles triangle A E B, and parallel to A B draw G H, equal to the proposed height. Then from E draw lines from the horizontal divisions of the cornice on A B, and the points in which these cut the line G H give the new heights. To find the projections, draw at the right side of the figure any horizontal line, C D, and on it draw perpendiculars from the projections of the various members of the cornice; produce the extreme perpendicular indefinitely; make D F equal to the perpendicular height of the isosceles triangle A E B, and from the divisions on C D draw lines to the point F. On F D set off F K equal to the perpendicular height of the isosceles triangle G E H, and draw I K parallel to C D. The divisions on I K, transferred to L M, on the left of the figure, give the projections diminished in the same ratio as the heights.

Let it now be required to enlarge the cornice at the left side of this figure. From the point G, with the proposed height of the cornice in the compasses, cut the line H N in N, and join G N: the points where this line is intersected by the horizontal lines of the mouldings are the heights of the members, enlarged in the same ratio as the whole height. To find the projections, from the point L draw the line L O, making the angle M L O equal to the angle H G N, and it will be cut by the lines of projections produced, in the same ratio as G N is by the horizontal lines. In place of drawing L O, the same result would be obtained by drawing a line at right angles to G N, crossing the lines of projection.

Fig. 2 is a further exemplification of the manner of enlarging and diminishing mouldings. First, as to enlarging:—From the point A, with the proposed increased height in the compasses, cut B O in O and join O A, and this line will be divided by horizontal lines drawn from any points, as *a*, *b*, in the mouldings, in the proportion that O A bears to B A. Then to find the projections, draw E P, making the angle D E P equal to the angle B A O, and vertical lines drawn from the same points in the mouldings as the horizontal lines will give the corresponding increased projections on P E.

Next, as to diminishing:—On A B construct an isosceles triangle A C B, and draw to C radial lines from the points of intersection of the horizontal divisions with A B. Draw I K parallel to A B, and equal to the proposed diminished height; then, to find the diminished projections corresponding to the divisions on I K, construct on D E an isosceles triangle D F E, having its vertical height equal to the vertical height of the triangle A C B on A B. To F draw radial lines from the divisions produced by perpendiculars drawn on D E from the points of projection, and intersect these by G H drawn parallel to D E, and at the same distance from it as I K is from A B. The divisions on G H, transferred to L M, at the right side of the figure, are the diminished projections.

Fig. 3 shows the manner of finding the proportions of a small moulding which is required to mitre with a larger one, or *vice versâ*. Let A B be the breadth of the larger moulding, and A D that of the smaller one; construct with these dimensions the parallelogram A D C B, and draw its diagonal A C; draw parallel to B C lines *a s*, *b t*, &c., meeting the diagonal in *s t*, &c., and from these points draw parallels to A B, meeting A D in *n*, *o*, *p*, *r*, &c.; produce them to *i*, *k*, *l*, *m*, &c., and make *n i* equal to *e a*, *o k* to *f b*, &c., and thus complete the contour of the moulding on D A, the lengths of which are diminished in the ratio of A D to A B, but its projections are as those of the larger moulding. The operation may be reversed, and the larger produced from the smaller moulding.

PLATE XC

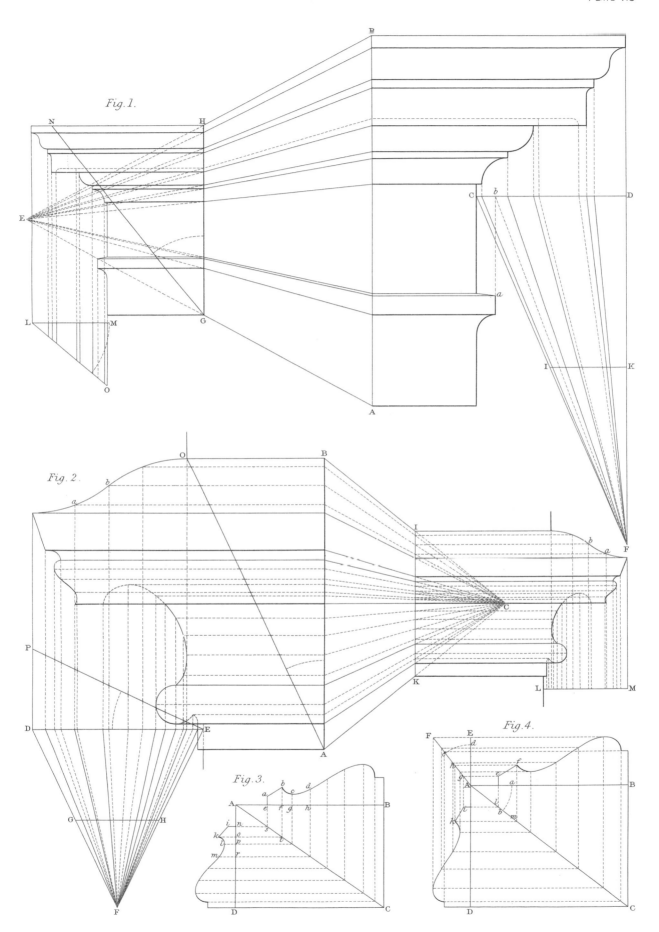

Fig.1.

Fig.2.

Fig.3.

Fig.4.

ENLARGING AND DIMINISHING MOULDINGS

Fig. 4 shows the manner of enlarging or diminishing a single moulding. Let A B be a moulding which it is required to reduce to A D. Make the sides A B, D C, and A D, B C, of the parallelogram respectively equal to the larger and smaller moulding, and draw the diagonal A C, produce D A to E, and make the angle E A F equal to *a* A *b*, and draw A F. The manner of obtaining the lengths and projections with these data is obvious.

CHAPTER II

JOINTS

Joinery is the art of cutting out, dressing, uniting, and framing wood for the external and internal finishings of buildings. It has been broadly distinguished from carpentry by this, that while the work of the carpenter cannot be removed without affecting the stability of a structure, the work of the joiner may. The labours of the carpenter give strength to a building, those of a joiner render it fit for habitation.

In joinery the parts are nicely adjusted, and the surfaces exhibited to the eye are carefully smoothed. The quality of the work depends, first, on the nature and seasoning of the materials, and, second, on the skill and care of the designer and workman. All the surfaces must be perfectly out of winding and smooth. The stuff must be square, the mouldings true and regular, and all must be framed together and fixed so as to be strong and durable. Moving parts, such as doors and windows, must work with ease and freedom. In many cases the wood must be cut in a special manner to bring out the beauty of the grain.

Many of the operations described in carpentry and joinery appear to be common to both, and such as may be performed indifferently by the carpenter or the joiner; but in reality it is not so, for a man may be a competent carpenter without being a joiner at all; and although a joiner may be able to execute carpentry work, yet the habits of greater neatness and precision required in the practice of his own proper art make his services less profitable in carpentry work.

The timber on which the joiner operates is termed *stuff*, and consists of boards, planks, deals, and battens. A "board" is a piece of timber usually $1\frac{1}{2}$ inch or less in thickness and wide in proportion. The term "plank" is generally applied to pieces 11 or more inches wide and not more than 4 inches thick. A "deal" is 9 inches wide and about 3 inches thick, and a "batten" from 2 to 7 inches wide. These terms are used in a more restricted sense than this in some places, but the definition applies generally.

Mortise-and-tenon Joints.—Mortising and tenoning in joinery are similar to the same operations in carpentry, with the exception of the variation arising from the smallness and neatness of the work. The parts should fit easily together without hard driving, and the tenon should fill the mortise fully and equally. The tenon is generally about one-fourth or one-third of the thickness of the framing, and its width seldom more than five times its thickness. If, therefore, the piece of wood to be tenoned is very wide, a double tenon should be formed, as at B and C, fig. 1117. The term "double tenon" is also applied, and more properly, to two tenons side by side, as in fig. 1118; this is suitable for thick doors, and for lock-rails, where mortise-locks are to be fixed at the level of the rails; obviously a single tenon would be cut away to receive the lock.

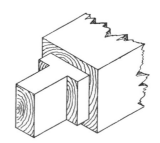

Fig. 1087.—Stump Tenon

If the piece of wood is also thick as well as wide, two projections, resembling short tenons, and called *stump-tenons* (fig. 1087), are made, one on each side of the main tenon, and fit into corresponding recesses in the mortised piece.

The haunched tenon has a small projection on one edge of the tenon, as shown at A in fig. 1117; it is used for the rails of doors, the haunch filling the end of the panel-groove in the stile.

The housed tenon (fig. 1088) is sometimes adopted where a narrow piece of wood is framed into a wider piece. The tenon is of the ordinary shape, but a recess about $\frac{1}{2}$ inch deep is formed around the mortise to receive the uncut portion of the tenoned piece.

The barefaced tenon has shoulders on three sides only (B, fig. 1113), and is used for the framing of framed and battened doors, the boards keeping the tenon in position on the fourth side.

As the stuff on which the joiner operates is of limited width, and it is frequently necessary to cover large surfaces, recourse is had to various modes of joining the pieces laterally. In these joints several expedients, as circumstances require, are used for the purpose of preventing air or dust blowing through, and also preventing the almost inevitable shrinkage of the timber from injuring the appearance of the work. These lateral joints may be dowelled, grooved and tongued, or rebated.

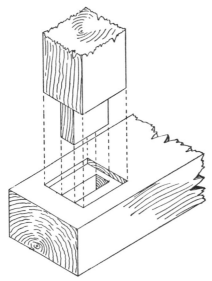

Fig. 1088.—Housed Tenon

Dowelling consists in forming corresponding holes in the contiguous edges of the boards, into which cylindrical wooden or iron pins are inserted, as in fig. 1089. The edges of the boards must be planed perfectly true, the operation being known as *shooting*.

Grooving and tonguing, or *ploughing and feathering* (fig. 1090), consists in forming a groove or channel along the edge of one board, and a projection or tongue to fit it on the edge of the other board. When a series of boards has to be joined, each board has a groove on one edge, and a feather or tongue on the other. These two last joints are commonly used for floors. The first is used without the dowels in common folded floors. The shrink-

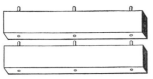

Fig. 1089.—Dowelled Joint

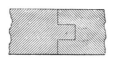

Fig. 1090.—Tongue-and-groove Joint

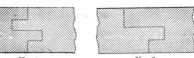

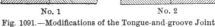

Fig. 1091.—Modifications of the Tongue-and-groove Joint

ing of the boards in this case causes the joint to open, and the air and dust pass through. The grooved and tongued joint is used in the better kind of floors; the tongue or feather prevents the passage of air or dust.

No. 1, fig. 1091, is a double-tongued or feathered joint. No. 2 is a combination of a rebate with a groove and tongue. It affords in flooring a better means of nailing, known as *secret-nailing*, as the nail-heads are hidden by the tongue and rebate of the right-hand board. The modification shown in No. 3, fig. 810, is also well adapted for secret-nailing.

When the boards are thick, grooves are made on the contiguous edges of both boards, and a small *slip-feather* or *tongue*, generally of hard wood, is inserted into both. This method involves a little more labour than the ordinary ploughing and tonguing, but this is more than compensated, in the case of thick stuff, by the saving in material. Examples are given in fig. 1092: No. 1 has a single softwood slip-feather, No. 2 two hardwood slip-feathers, and No. 3 a single tongue of wrought-iron; No. 4 shows dovetail grooves, with a slip-feather of corresponding form, which must be inserted endways. This variety cannot, of course, be used for floors.

The grooved and tongued joint is also used for fastening the ends of boards together by means of another board at right angles, as shown in fig. 1093. The tongues are formed on

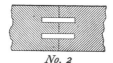
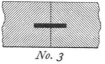
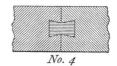
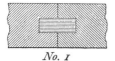

No. 1 No. 2 No. 3 No. 4

Fig. 1092.—Grooved Joints with loose Tongues or Slip-feathers

the ends of the boards, and the groove is formed in the cross-piece. This is known as *clamping.* To avoid exposing the end grain of the cross-piece at A, *mitre-clamping* may be adopted, as at B. Sometimes all the pieces are grooved and a slip-feather inserted.

Fig. 1094 is a simple rebated joint. One half of the thickness of each board is cut away to the same extent, and when the edges are overlapped the surfaces lie in the same plane.

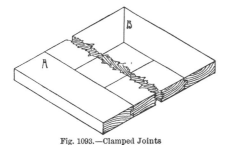

Fig. 1093.—Clamped Joints

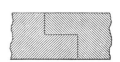

Fig. 1094.—Rebated Joint

Where a large surface has to be covered with boarding not framed, the deals are cut into narrow widths, and joined at their edges by some of the joints just described. The shrinkage of the wood often causes the simple groove-and-tongue joint to open, and thus disfigures the work. To prevent this disfigurement a small moulding, termed a *bead*, is sometimes run on the edge of each board, as in No. 1, fig. 1095. The joint thus forms one of the quirks of the bead, and prevents any slight opening from the shrinkage of the wood being observed. This is termed a grooved, tongued, and beaded joint. So also in the case of the rebated joint, a bead may be run on the edges of the board, and the result is as in No. 2. This is termed a rebated and beaded joint. Instead of being beaded,

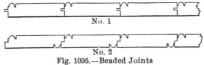

Fig. 1095.—Beaded Joints

No. 1, Tongued and grooved; No. 2, rebated

the edges of each board may be slightly chamfered, so that, when the boards are fixed, a V-shaped channel is formed along each joint; these are known as V-jointed boards, and may of course be also tongued and grooved or rebated. All boards with the edges made to joint with corresponding boards (such as those shown in figs. 1090, 1091, 1094, and 1095) are known as "match boarding".

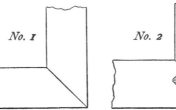
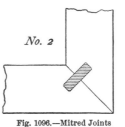
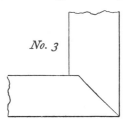

No. 1 No. 2 No. 3

Fig. 1096.—Mitred Joints

Angle Joints.—In joining angles formed by the meeting of two boards various joints are used, among which are those which follow.

In fig. 1096, No. 1 is the common mitre-joint, used in joining two boards at right angles to each other. Each edge is planed to an angle of 45°. No. 2 shows a mitre-joint keyed by a slip-feather, and No. 3 a mitre-joint when the boards are of different thickness. The mitre on the thicker piece is only formed to the same extent as that on the thinner piece; hence there is a combination of the mitre and simple butt joint.

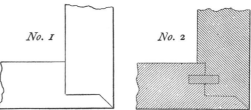

Fig. 1097.—Rebated and Mitred Joints

Fig. 1097 shows other methods of joining two boards of either the same or of different

thicknesses.   In No. 1 one of the boards is rebated, and only a small portion at the angle of each board is mitred.   In No. 2 both boards are rebated, and a slip-feather is inserted as a key.   These joints may be nailed through from both faces.

Fig. 1098 shows combinations of grooving and tonguing with the last-described modes. These can be fitted with great accuracy and joined with certainty.   No. 3 is a joint formed

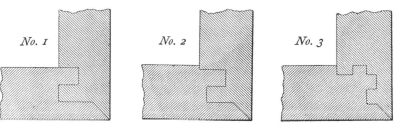

by the combination of mitring with double grooving and tonguing.   The boards must in this case be slipped together endways, and cannot be separated by a force applied at right angles to the planes of their surfaces.

Fig. 1098.—Tongued, Grooved, and Mitred Joints

In the butt-joints which follow, the edge of the one board abuts against the face of the other, the edge of which is consequently in the plane of the surface of the first board, the shrinkage of which would cause an opening at the joint.   To make this opening less apparent is the object of forming

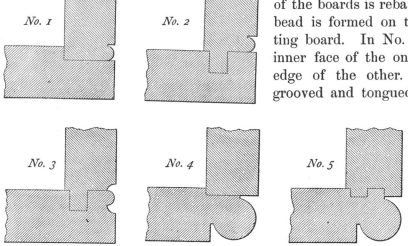

the bead-moulding seen in fig. 1099.   In No. 1 one of the boards is rebated from the face, and a small bead is formed on the external angle of the abutting board.   In No. 2 a groove is formed in the inner face of the one board and a tongue on the edge of the other.   In No. 3 the boards are grooved and tongued as in the last figure; and a cavetto is run on the external angle of the abutting board, and the bead and a cavetto on the internal angle of the other board.   In No. 4 a quirked bead is run on the edge of one board, and the edge of the abutting board forms the double quirk.   In No. 5 a quirked

Fig. 1099.—Beaded Joints for right-angles

bead is formed at the external angle, and the boards are grooved and tongued.   The external bead is attended with this advantage, that it is not so liable to injury as the sharp arris.   There are many positions, however, in which it cannot well be used, such as the angles of pilasters and wall-panelling.

Fig. 1100 shows the same kind of joints as have been described, applied to the framing together of boards meeting in an obtuse angle.

In fig. 1101 joints used in putting together cisterns are shown.   Nos. 3 and 4 are of the dovetail form, and require to be slipped together endways.   For small cisterns lined with lead or copper the angles are often dovetailed, as in fig. 1103 (No. 1).

Nos. 1 and 2, fig. 1102, show methods of joining boards together laterally by keys, in the manner of scarfing; and No. 3 shows another method of securing two pieces, such as those of a circular window frame-head, by keys.   In the latter a shallow mortise is cut in each piece and a hardwood key of the shape shown is placed in the mortises, and the joint made tight by the wedges indicated by the shaded portions.   The term "keying" is also applied to the operation of fixing a number of boards together in the same plane by means of a tapering board driven into a dovetail-shaped chase sunk in the other boards.

Dovetail-joint.—This joint (fig. 1103) has three varieties.—1st, the common dovetail (No. 1), where the dovetails are seen on each side of the angle alternately; 2nd, the lapped

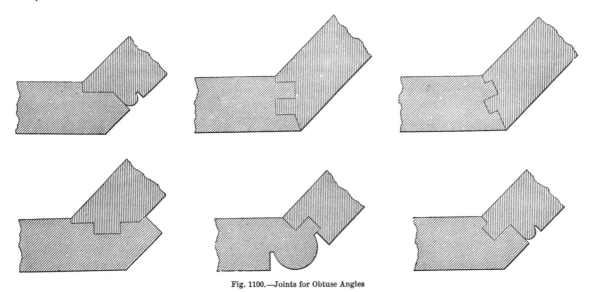

Fig. 1100.—Joints for Obtuse Angles

dovetail (No. 2), in which the dovetails are seen only on one side of the angle; and 3rd, the lapped and mitred dovetail (No. 3), in which the joint appears externally as a common

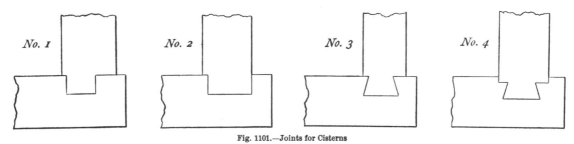

Fig. 1101.—Joints for Cisterns

mitre-joint. The lapped and mitred joint is useful in salient angles, in finished work, but it is not so strong as the common dovetail, and therefore in all re-entrant angles the latter should be used. In No. 1, A is an elevation of the common dovetail-joint; B, a perspective representation; and C, a plan of the same. The pins or dovetails of the one side are marked A, and those of the other side are marked B. No. 2, D, E, F.—In these the lapped dovetail-joint is represented in plan, elevation, and perspective projection. No. 3, G, H, I, K.—In these figures the mitred dovetail-joint is represented in plan, elevations, and perspective projection. The dovetails of the adjoining sides are marked respectively B and C in all the figures.

Nos. 1 and 2, fig. 1104, show the modes of dovetailing an angle when the sides are inclined to the horizon, as in a hopper. The pins of the one side are marked A, and those of the other side B.

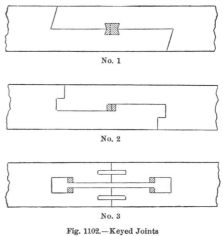

Fig. 1102.—Keyed Joints

In fig. 1105 various methods of jointing for architraves are shown. In No. 1 the two pieces are united by a rebate-and-groove joint, in No. 2 by a tongue-and-groove joint, and in No. 3 by a slip-feather. The joint in No. 1 is better than that in No. 2, as in the latter any

opening at the joint, due to shrinkage, will be seen. The faint lines outside the shaded portions show the shaped blocks on which the architraves stand.

The modes of joining timber above described are all more or less imperfect. The liability of wood to shrink renders it essential that the joiner should use it in such narrow widths as to prevent this tendency marring the appearance of his work; and, as even when so used it will still expand and contract, provision should be made to admit of this. The

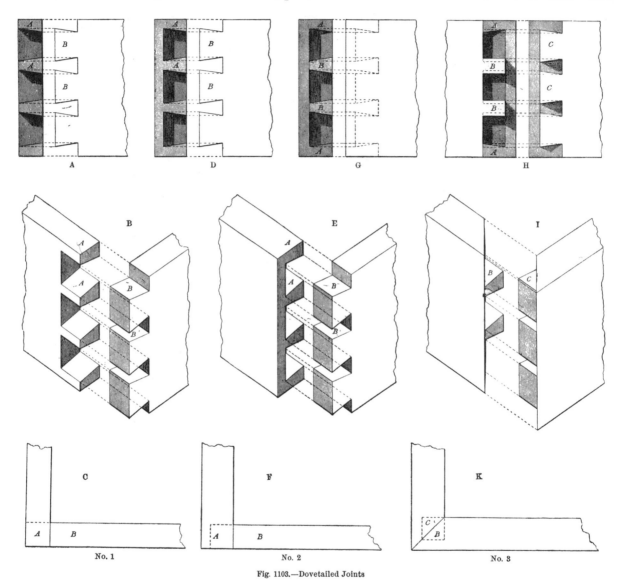

Fig. 1103.—Dovetailed Joints

groove-and-tongue joint admits of a certain amount of variation, and the grooved, tongued, and beaded joint admits of this variation with a degree of concealment, but the most perfect mode of satisfying both conditions is by the use of framed work.

Framing in joinery consists of pieces of wood (usually of the same thickness) fixed together so as to inclose a space or spaces. These spaces are filled in with boards of a less thickness, termed *panels*.

In fig. 1106 *a a*, *b b* shows the framing, and c c the panels. The vertical pieces of the framing are termed *stiles*, and the horizontal pieces are termed *rails*. The rails have tenons which are let into mortises in the stiles. The inner edges of the stiles and rails are grooved to receive the edges of the panels, and thus the panel is at liberty to expand and contract. Framing is always used for the better description of work. Wide panels (par-

ticularly those of deal and other softwoods) should be formed of narrow pieces glued together, with the grain reversed alternately.

The panels may be boards of equal thickness throughout, in which case the grooves in the stiles and rails are made of sufficient width to admit their edges, as in fig. 1107, No. 1. These are termed *flat* panels. In this example the framing is not moulded, and is termed *square*. In No. 2 the framing is moulded on the solid on one side, and in No. 3 mouldings are planted on both sides; these mouldings project beyond the face of the framing, and are known as *bolection mouldings*.

Flush panels have one of their faces in the same plane as the face of the framing, and are rebated round the edges until a tongue sufficient to fit the groove is left. Flush panel framing has generally a simple bead stuck on its edges all round the panel, and the work is called *bead flush* (No. 4). In No. 5 the framing is bead flush on one side, and a moulding is planted on the other side. In inferior work the bead is run on the edges of the panels in the direction of the grain only, that is, on the two sides of each panel, while its two ends are left plain (No. 6); this is termed *bead butt*.

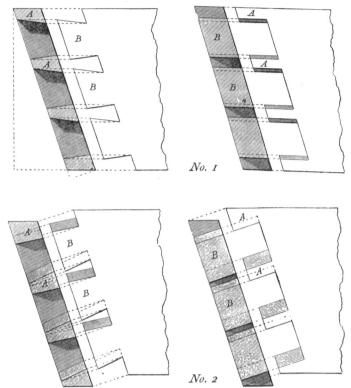

No. 1

No. 2

Fig. 1104.—Raking Dovetailed Joints

Raised panels (Nos. 7 and 8) are those of which the thickness is such that one of their surfaces is a little below the framing, but at a certain distance from the inner edge, all round, begins to diminish in thickness to the edge, which is thinned off to enter the groove. The line at which the diminution takes place is usually marked either by a square sinking or a moulding.

Panels in external work, such as doors, may be secured against being cut through by depredators by boring holes across them from edge to edge and inserting iron wires, or by crossing them diamond fashion with thin hoop iron nailed on the inside.

In fig. 1108 a variety of mouldings for framed work is given to a larger scale, and the use of them will now be better understood. No. 1 shows a bead-flush panel on one side, and a planted moulding on the other. Nos. 2, 3, and 4 are moulded on the solid on both sides, the panel in No. 2 being flat, in No. 3 raised on one side, and in No. 4 raised on both sides.

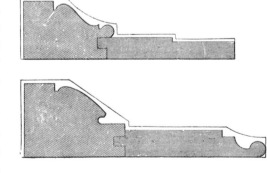

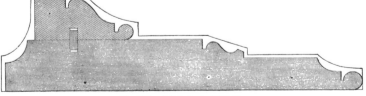

Fig. 1105.—Joints in Architraves

No. 5 shows two simple forms of planted mouldings. Nos. 6, 7, and 8 are bolection-moulded, No. 9 is also bolection-moulded, but the groove for the panel is formed in a separate piece which is housed into the framing. In No. 10 the moulding itself is grooved to receive the panel, and is mitred around the framing and fixed to it by means of two tongues and grooves. No. 11 is a combination of Nos. 9 and 10.

The examples which have been given do not by any means exhaust the subject of joints, but are those most generally used by the joiner. Special joints are, however, required in many cases, and of these many examples will be found in Section X, Part I, and Section XI, and others will be given in the following chapters.

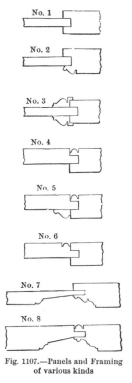

Fig. 1107.—Panels and Framing of various kinds

Gluing up Columns, &c.—PLATE XCI.—Fig. 1, Nos. 1 to 5, show in detail the manner of framing together and gluing up the parts of a column and its entablature.

Fig. 1, No. 4, contains four quarter plans of as many courses of timber, forming the mouldings of the base; and No. 5 shows the same in elevation and section, marked with the same letters. The square plinth is first formed by taking four pieces of equal length, mitring them together at the angles, and securing them by screws and by glue, and strengthening them with blockings in the angles, as at *u*. The upper surface of this course is then planed true, and prepared to receive the next course, which is also constructed of four pieces, to form the torus, A, No. 5. These pieces cross the angles, and are joined on the middle of the length of the first pieces. The surface of this course is in like manner planed true to receive the next course B; the pieces composing which have their joints on the middle of the pieces of the course below, and are glued down on

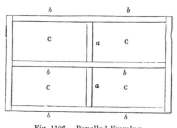

Fig. 1106.—Panelled Framing

them. So likewise with c. When the formation of the block is thus completed, it is turned, and the proper rebate *e* formed in the upper surface to receive the lower part of the shaft. The quadrant marked D shows a section of the lower part of the shaft. The half-plan at o o, on the right hand of fig. 1, No. 3, is the upper part of the shaft immediately below the necking (A B, No. 1). It will be observed that the shaft is built of staves, which should always have their joints in the centre of the fillet. The staves are glued together and secured by blockings *m m m*, glued in the angles. The staves should not exceed 5 inches wide, whether for columns or pilasters, and they should be as thin as is consistent with strength. It is well to work them to the taper necessary for the diminution of the columns before gluing them together.

The quadrant at R, No. 3, is part of the horizontal section at C D, No. 1, and s a quadrant of the horizontal section at E E.

No. 1, fig. 1, is the elevation of half the capital, and the architrave, frieze, and cornice, and No. 2 is a vertical section of the same. The framing of the architrave, frieze, and cornice does not require description; by using blockings the thickness of the material in the architrave and frieze might be reduced. Some methods of building up smaller cornices have been given in fig. 1039.

Diminution of Columns—Fig. 2.—The upper diameter of a classic column is less than its lower diameter, but the gradual diminution between them is not made by straight but by curved lines. The usual mode of describing the curved contour of the diminution is as follows:—Let *a b* be equal to the lower diameter of the column, of which let *e f g* be the line of the axis perpendicular to *a b*, *f g* the height of the column, and *r* 6 its upper diameter.

PLATE XCI

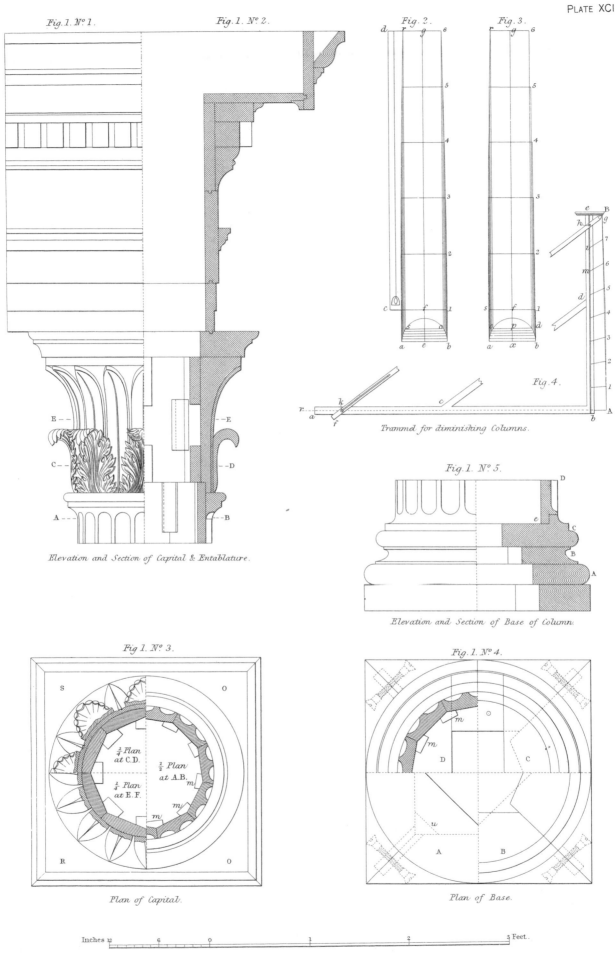

Fig. 1. Nº 1.   Fig. 1. Nº 2.

Fig. 2.   Fig. 3.

Fig. 4.

Trammel for diminishing Columns.

E - - - E

C - - - - D

A - - - - B

Elevation and Section of Capital & Entablature.

Fig. 1. Nº 5.

Elevation and Section of Base of Column.

Fig. 1. Nº 3.

¼ Plan
at C.D.

½ Plan
at A.B.

¼ Plan
at E.F.

Plan of Capital.

Fig. 1. Nº 4.

Plan of Base.

Inches 12   6   0   1   2   3 Feet.

GLUING UP COLUMNS

On $a\,b$ describe a semicircle, and from r and 6 draw lines parallel to the axis, cutting the semicircle in $s\,o$; divide $s\,a$ or $o\,b$ into any number of equal parts, the more the better, and divide the height $f\,g$ into the same number of equal parts, as 1, 2, 3, 4, 5, 6, and through these draw lines crossing the axis perpendicularly. Then by drawing lines parallel to the axis through the corresponding divisions in the semicircle meeting these points, the curved

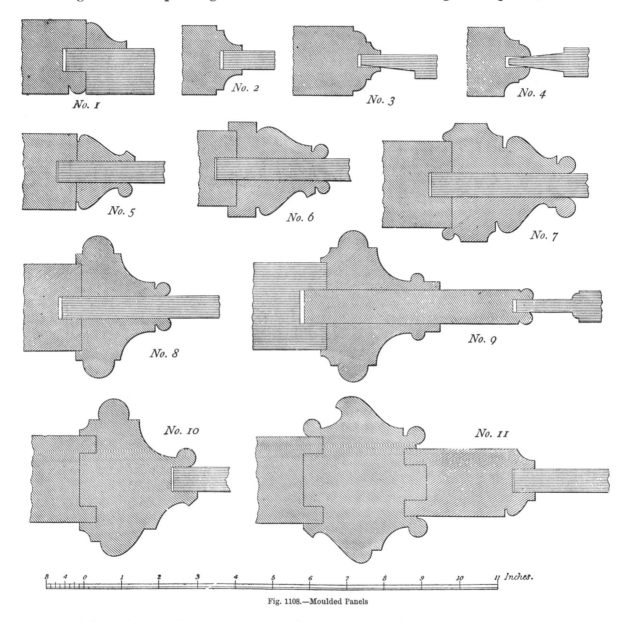

Fig. 1108.—Moulded Panels

contour of the column will be obtained, and by bending a lath, so as to pass through these points, the curve may be drawn, and the rule $c\,d$ formed.

Fig. 3.—The same result may be obtained in a manner somewhat different, as shown in fig. 3. In this $a\,b$ is equal to the lower, and $c\,d$ to the upper diameter. The points in which this latter cuts the semicircle being found, the portion of the radius $x\,p$ is divided into certain equal parts, and the height of the column $f\,g$ into the same number of equal parts, and from the points where lines parallel to $a\,b$, drawn through the divisions in $x\,p$, meet the semicircle, other lines parallel to the axis are drawn, as before, to intersect the lines drawn through the divisions of the height 1, 2, 3, 4, 5, 6.

Another method of describing the vertical section of the column is shown in fig. 4. Let $b\,e$ be the line of the axis of the column, A b half of the lower diameter, and B e half of the

upper diameter. Take in the compasses the length of the semi-diameter at the bottom, and setting one foot in the extremity of the upper diameter at B, with the other foot cross the axis at *h*, produce the lower diameter indefinitely, as A *r*; and through B, and the point *h* on the axis, draw a line cutting the line A *r* in *k*; then from *k* as a centre draw any number of lines, as *i* 7, *m* 6, &c., and make each of them, as *i* 7, equal to the lower semi-diameter. In the same figure is represented a trammel for doing the same thing as has been described: *a b e* is a right-angled rule, kept to its form by the angle-piece *c d*. In the limb *b e* is a groove, which is made to coincide with the axis of the column, and in which slides freely a stud *h*. The other arm *a b* of the rule carries a stud *k*. The rule *f g* has a groove or slot sliding on the stud *k*, and its other end carries the stud which slides in *b e*. It is evident that if the points *k*, *b*, *h*, *g* of the trammel are adjusted in accordance with the preceding description, the point *g* will, on the rule *f g* being slid along, guided by the grooves, describe the elliptic curve A, 1, 2, 3, 4, 5, 6, 7, *g*.

CHAPTER III

DOORS

Doors are of different kinds, and are classified according to their construction and method of hanging. The simplest form is the ledged door; the ledged and braced door is slightly more complex, and after this come the framed and ledged door, the framed, ledged, and braced door, and the panelled door with or without glazed panels. Classified according to the method of hanging, doors may be either single, folding, swing, sliding, or revolving.

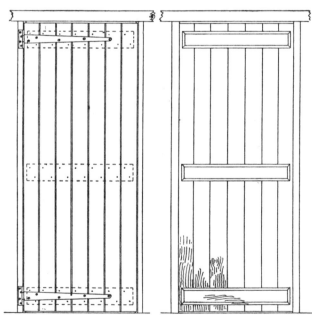

Fig. 1109.—Face and Back of Ledged Door

Doors in horizontal planes, such as floors and ceilings, or in inclined planes, such as roof-slopes, are known as trap-doors. Fire-resisting doors are now made of double or treble boarding covered with sheet metal, and a similar method of construction (without the metal covering) is often adopted for ordinary doors, moulded pieces being planted on to give the desired relief.

The ledged door (fig. 1109) is used for cellars, out-buildings, and other places where more elaborate doors are unnecessary or too costly. It consists of a number of vertical boards (generally ploughed and tongued, and beaded or chamfered at the joints) nailed to three horizontal battens or boards known as the top, middle, and bottom ledges. The vertical boards are usually $\frac{3}{4}$ inch or 1 inch in thickness, and from $3\frac{1}{2}$ to 5 inches wide, and the ledges 1 or $1\frac{1}{4}$ inches thick, and about the same width as the boards. The boards are secured to the battens with wrought-iron nails, driven from the face of the boarding; sometimes the nails are $\frac{1}{4}$ or $\frac{1}{2}$ inch longer than the combined thickness of the boarding and batten, and are clenched on the inside to prevent them from working out. Doors of this kind seldom exceed 2 feet 9 inches in width, unless they are provided for temporary purposes.

Ledged doors are usually hung with iron hinges, known as "cross-garnets" or "T-hinges", which are fixed with screws or small bolts. If the door opens outwards, the hinges are fixed on the face side, and the frame is rebated outside to the thickness of the boarding. If the door opens inwards, the hinges may be fixed to the back of the boarding, the frame being rebated inside to the same depth as in the first case; or the hinges may be fixed to the backs of the top and bottom ledges, the frame being either rebated to the

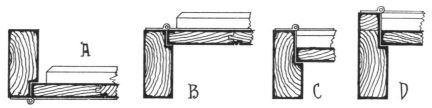

Fig. 1110.—Methods of hanging Ledged Doors

combined thickness of the boarding and ledge, or rebated to the thickness of the boarding, and blocked out with wood equal in thickness to the ledge. These methods of hanging are shown at A, B, C, and D, fig. 1110; if the hinges are fixed to the ledges, these must be extended to the edge of the door in order to prevent undue strain on the hinges.

The ledged and braced door (fig. 1111) differs from the ledged door in having oblique battens on the back between the ledges. These oblique members, although known as "braces", are in reality "struts", designed to transmit the weight of the door to the inner ends of the middle and bottom ledges, and so prevent the dropping of the nose or free end of the door. The ends of the braces are cut to an angular shape, and fit into corresponding notches in the ledges.

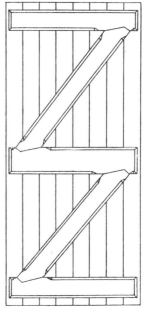

Fig. 1111.—Back of Ledged and Braced Door

Ledged and braced doors of large size are often used in farm-buildings, and are hung with wrought-iron band hinges about 2 inches wide and $\frac{5}{16}$ inch or $\frac{3}{8}$ inch thick to "hooks" or "crooks", which are either built into the brickwork or secured to "hook stones" by means of wedge-shaped holes caulked with lead. When this method of hanging is adopted, a wood frame is not required, but a chamfered facing $\frac{3}{4}$ inch thick is often nailed to the face of the door, and scribed to the brick or stone reveal, so as to form a fairly weather-tight joint. A similar facing is also used to cover the meeting of the two leaves, when the doors are hung folding. A pair of doors of this kind is shown in fig. 1112. Two of the ledges in each leaf are continued beyond the free edge of the leaf, and a wood bar is pivoted to the lower of the two ledges in such a manner that, when it is turned to an inclined position, it clips one or two ledges of the other leaf; when the corresponding bar on the other leaf is also moved into position, the doors are effectually barred. The upper part of each bar moves in an iron staple fixed to the higher of the two ledges.

The boards for large doors of this kind ought to be 1 or $1\frac{1}{4}$ inch thick, and the ledges $1\frac{1}{4}$ or $1\frac{1}{2}$ inch thick, and from $3\frac{1}{2}$ to $5\frac{1}{2}$ inches wide.

Stable doors are usually made in two heights, so that the upper portion can be left open for ventilation while the lower portion is closed; they should be about 4 feet wide. Cow-house doors may be 3 feet 6 inches wide.

The framed and ledged door (fig. 1113), known also as the *bound door*, is stronger and more costly than the ledged door. The total thickness ought not to be less than what is known as 2 inches. The framework of such a door will consist of two stiles 2 inches thick and about $4\frac{1}{2}$ inches wide, a top rail of the same dimensions tenoned into them, and middle

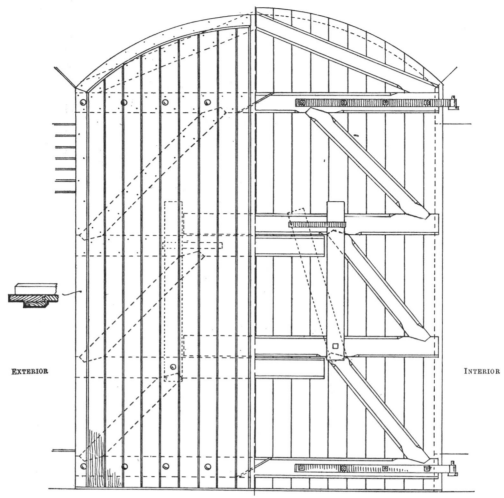

Fig. 1112.—Face and Back of Ledged and Braced Folding Doors

and bottom rails about 9 inches wide, and equal in thickness to the difference between the thicknesses of the boarding and the main framing; if the boarding of a 2-inch door is 1 inch thick, the middle and bottom rails must also be 1 inch thick, but if $\frac{3}{4}$-inch boards are used, the thickness of these rails must be $1\frac{1}{4}$ inch. Barefaced tenons are used for fixing the ends of the middle and bottom rails to the stiles, as shown in the enlarged view at B. The top rail and stiles are ploughed, and the boards rebated to fit into the grooves, as shown in the detail at A; the boards run over the middle and bottom rails, and are nailed to them from the face. The boards may be either vertical or diagonal. Sometimes the bottom rail is the same thickness as the stiles, but this forms a horizontal joint in which water may lodge and lead to decay in the wood.

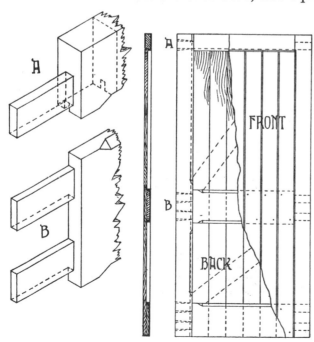

Fig. 1113.—Framed and Ledged Door

If diagonal "braces" are inserted, as shown by the dotted lines, the door is known as a *framed, ledged, and braced door*.

Doors of this kind may be hung with wrought-iron band hinges to hooks, or with

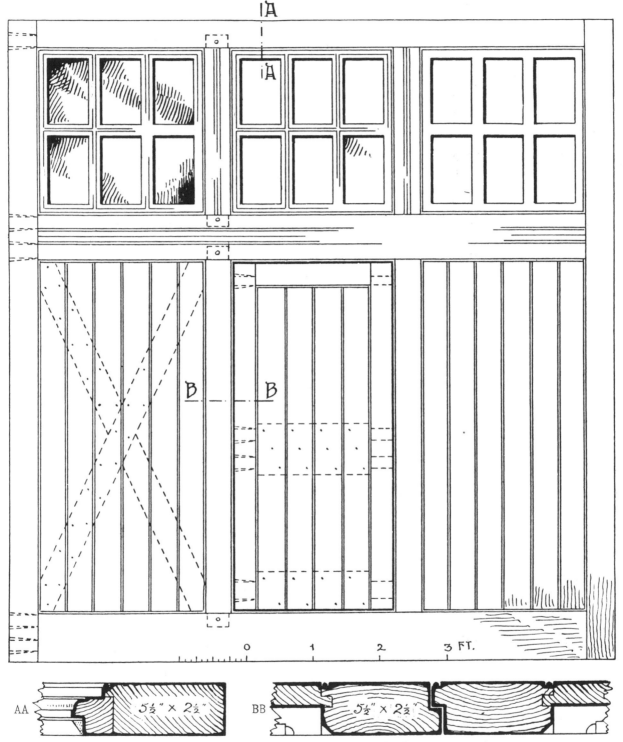

Fig. 1114.—Large Framed, Ledged, and Braced Sliding Doors

strong butt hinges, or plain or ornamental strap hinges to rebated wood frames. For large framed doors an iron pivot, working in a gun-metal socket fixed in the stone threshold, is sometimes used instead of the lower hinge-knuckle and hook. The door is not so likely to drop at the nose, if this method is adopted.

A more elaborate framed and ledged door is shown in fig. 1114. It is 9 feet wide, and is one of a pair, designed to slide, the one in front of the other. The frieze rail is moulded on the face, and the panels above are fitted with 2-inch glazed casements. A small door is formed in the central portion of the main door, and hung to one of the muntins; it ought to be hung in such a manner that, even when open, it will not interfere with the sliding of either of the main doors. The braces in the side portions are arranged diagonally, and are halved together at their intersections. Various methods of hanging sliding doors will be considered at a later stage.

Panelled doors are most frequently used and can be made in endless variety. They may be plain, chamfered, beaded, or moulded on one or both sides, and the panels may vary in number and shape, may be flush or sunk, plain or raised, or of glass or wood. The mouldings around the panels may be worked on the solid (technically, "stuck"), or may be worked separately and planted on. The heads of the doors may be square or arched.

In the framing of doors, as in other framing, the vertical pieces are termed *stiles* and the horizontal pieces *rails*. When an intermediate vertical piece is tenoned into mortises formed in the rails it is called a *muntin, montant,* or *mounting*. The rail next above the bottom rail is called the *lock* rail, and if another rail is placed between the lock rail and top rail, it is known as a *frieze* rail; other intermediate rails have no specific name. In like manner the panels are named *frieze* panels, *middle* panels, and *bottom* panels. In fig. 1115 *a a* are the stiles, *b* the muntin, *c* bottom rail, *d* lock rail, *e* frieze rail, *f* top rail, *g g* frieze panels, *h h* middle panels, *k k* bottom panels.

In ordinary framed doors the top and frieze rails are generally of the same width as the stiles, the bottom and lock rails generally twice as wide. The top of the lock rail is usually between 3 feet 2 inches and 3 feet 6 inches from the bottom of the door, if the lock is fixed at the level of the middle of the rail. If the lock is fixed entirely within the stile, the rail may be lower or higher according to the designer's fancy.

When a doorway is closed by two doors hinged to its opposite jambs, the middle or meeting stiles are frequently rebated and beaded; such a door is termed a *folding* or *two-leaved* door. Doors which, whilst they are in one width, are framed with a wide stile in the middle, beaded in the centre in imitation of the two stiles of a two-leaved door, are called *double-margined* doors. Fig. 1116 shows the appearance of the two-leaved and double-margined doors. A *sash* door is one which is glazed above the lock rail. A *jib* door is one which is flush with the surface of the wall of the apartment in which it is placed; it has no architraves or other ornamental border, but is crossed by the skirting, surbase, and other finishings of the apartment, and is otherwise so finished as to be (as nearly as possible) indistinguishable from the wall itself.

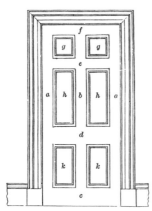

Fig. 1115.—Six-panelled Door

Fig. 1116.—A Pair of Folding Doors

The size usually adopted for internal doors in cottages measures 2 feet 6 inches wide and 6 feet 6 inches high, with a nominal thickness of $1\frac{1}{2}$ inch. For better houses larger doors should be used, but in this country the internal doors of middle-class houses do not often exceed 3 feet 3 inches by 7 feet by 2 inches. In mansions, large hotels, and public buildings, the doors of the principal rooms must be still larger.

The most ordinary form of panelled door is that with four oblong panels; a door of this kind, $1\frac{1}{2}$ inch thick, is shown in fig. 1117. Being without moulds, it is known as a "plain" or "square-framed" door.

The stiles must be cut somewhat longer than the height of the door, and the rails somewhat longer than the width. Each muntin must be about 4 inches longer than the adjacent panels, in order to allow for a tenon, 2 inches long, at each end. The framework must be planed to uniform thickness and finished true.

The stiles and top and bottom rails must be ploughed along one edge to receive the panels, and the lock rail and muntins along both edges. The bottom of the groove in the top rail forms the bottom of the tenons at the ends as shown at A; the depth of these tenons should not greatly exceed half the depth of the rail, as otherwise the part of the stile above the mortise may be sheared out in the process of "wedging up". A haunch must be left above the tenon to fill the panel-groove of the stile. The width of the tenon is usually one-third the thickness of the door, the thickness of the panel being the same.

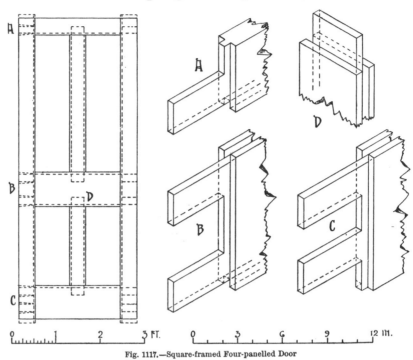

Fig. 1117.—Square-framed Four-panelled Door

The lock rail, being of greater depth, is usually formed with two tenons at each end, one above the other, as shown at B. The bottoms of the two panel-grooves in the rail are the top and bottom limits of the tenons.

The bottom rail must also have two tenons at each end as shown at C, but the bottom of the lower tenon should be at least 1½ inch from the finished bottom of the door to prevent damage in wedging up.

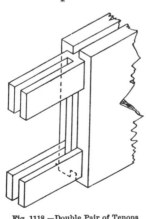

Fig. 1118.—Double Pair of Tenons at End of Lock Rail

The tenon at each end of a muntin may be equal in width to the width between the panel-grooves of the muntin, and about 2 inches long, as shown at D.

The mortises in the stiles must have the same width as the tenons, and the depth on the inner side must also equal the depth of the tenons. The top and bottom of the mortise must, however, slope upwards and downwards respectively to the outside of the stile to allow for the insertion of wedges. "Wedging up" is the last process in the framing of the door, and must not be done till the panels have been fitted into position, and the door has been well dried. The wedges must be somewhat blunt at the ends,

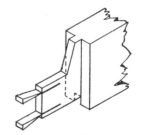

Fig. 1119.—Fox-tail Wedging for Top Rail of Door

so that they will compress the tenon at the root, and so prevent the tenon becoming loose by shrinkage, which will otherwise take place if the wood is not thoroughly seasoned.

Each panel may consist of a single board, or of two or more boards put together with plough-and-tongue joints (well glued) and afterwards planed to a true surface. The panels must not be nailed or wedged tightly to the framing, but must be free to move

at the edges, so that in the event of shrinkage they will not crack. If mouldings are planted around the panels, they must be nailed to the framing; if they are nailed to the panels, the latter may be cracked by shrinkage, or the mouldings may be drawn away from the framing.

For doors of a nominal thickness of 2 inches and upwards, the tenons at the ends of the rails should be double. The double tenon ought always to be made at the end of the lock rail, if a mortise lock is fixed at this level. The tenon at this point becomes a double pair, as shown in fig. 1118.

For hardwood doors, where it is not desired to expose the end grain of the tenons on the edges of the stiles, fox-tail wedging may be adopted. The mortises are cut into the stiles about 3 inches deep, and the tenons are slightly shorter, so that if the wedges fold over in driving, the rail will still come to its bearing against the stile. Fig. 1119 shows the joint at the end of the top rail. The wedges are inserted into saw-cuts at the end of the tenon, and the tenon is then placed in the mortise (which must be cut to a dove-tail shape), and driven till the shoulder fits closely against the stile.

Veneering is sometimes adopted for hardwood doors, but is more suitable for the panels than for the framing. In a new type of Canadian door, which is now being introduced into this country, hardwood facings are attached to the softwood door by means of solid tongues and grooves.

Doors may be framed and panelled in an almost

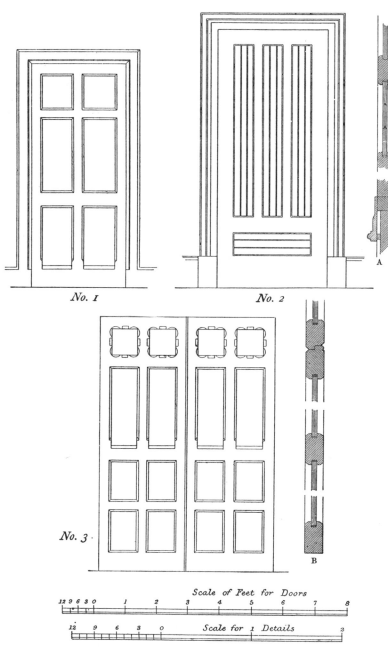

No. 1   *No. 2*

No. 3

Scale of Feet for Doors
12 9 6 3 0   1   2   3   4   5   6   7   8

12   9   6   3   0   Scale for 1 Details   2

Fig. 1120.—Three Panelled and Chamfered Doors

infinite variety of ways. Fig. 1120 contains three examples of simple panelled doors with chamfered framing. No. 1 has six panels, and the chamfers on the stiles and muntin are "stopped" at the sides of the bottom panels. In No. 2 the panels are formed with narrow match-boards, usually V-jointed; an enlarged horizontal section through part of the door and frame is shown at A; a stop *a* (usually ½ inch thick) is planted around the frame to form the rebate for the door, and the architrave is nailed to the frame and also to a rough ground

PLATE XCII

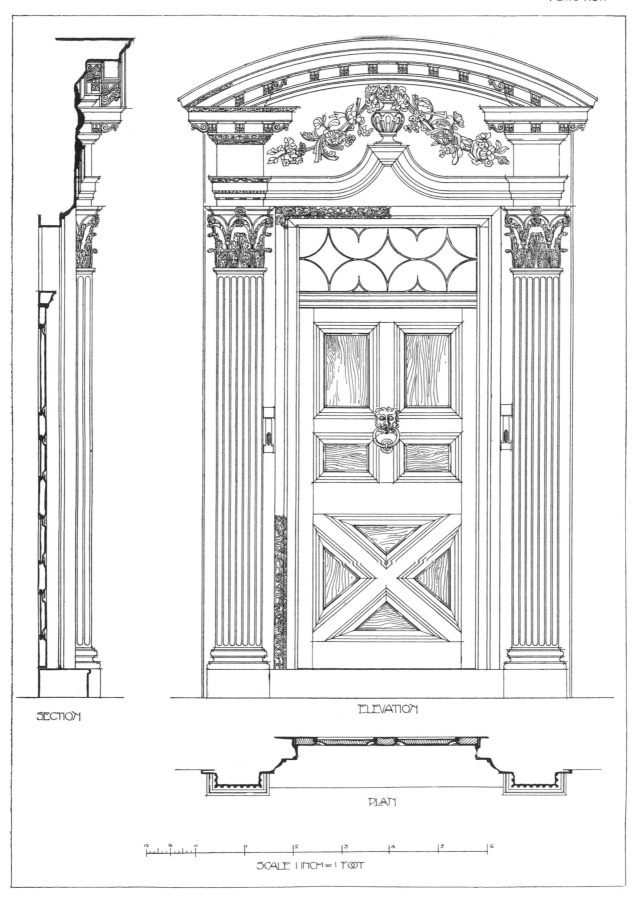

SECTION

ELEVATION

PLAN

SCALE 1 INCH = 1 FOOT

CARVED PINE DOOR FORMERLY IN A HOUSE IN LINCOLN'S INN FIELDS

(NOW IN SOUTH KENSINGTON MUSEUM)

PLATE XCIII

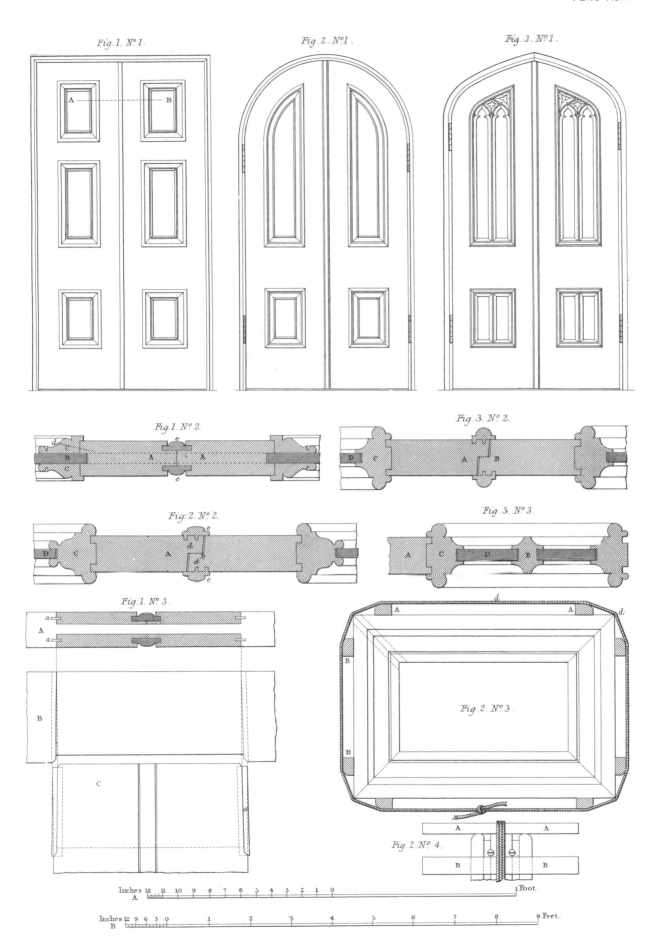

Fig. 1. Nº 1.   Fig. 2. Nº 1.   Fig. 3. Nº 1.

Fig. 1. Nº 2.   Fig. 3. Nº 2.

Fig. 2. Nº 2.   Fig. 3. Nº 3.

Fig. 1. Nº 3.

Fig. 2. Nº 3.

Fig. 2. Nº 4.

Inches 12 11 10 9 8 7 6 5 4 3 2 1 0   1 Foot.
A

Inches 12 9 6 3 0 1 2 3 4 5 6 7 8 9 Feet.
B

DOUBLE-MARGINED AND FOLDING DOORS

PLATE XCIV

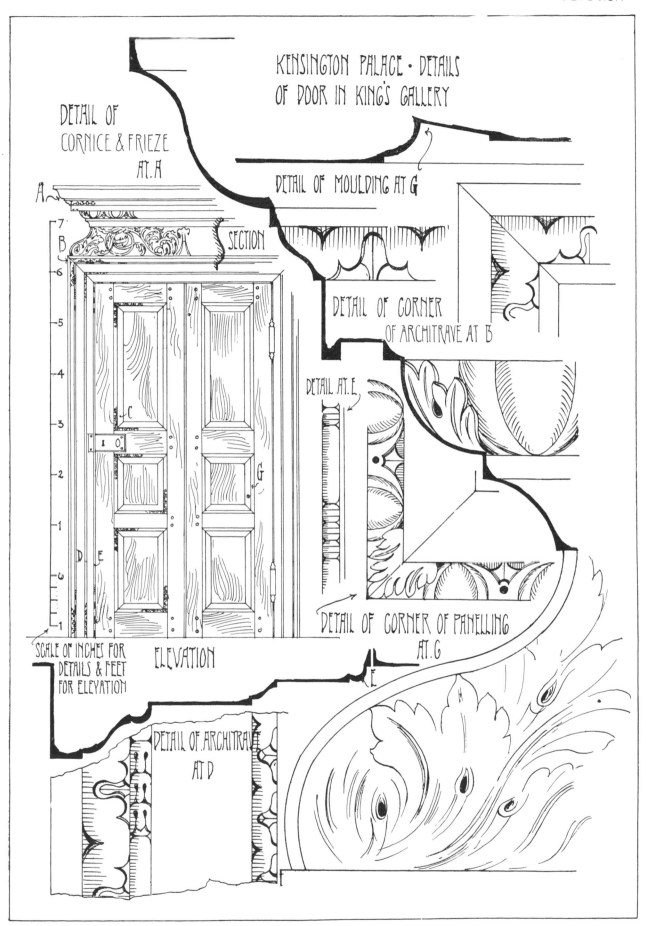

KENSINGTON PALACE · DETAILS OF DOOR IN KING'S GALLERY

DETAIL OF CORNICE & FRIEZE AT A

DETAIL OF MOULDING AT G

DETAIL OF CORNER OF ARCHITRAVE AT B

SECTION

DETAIL AT E

DETAIL OF CORNER OF PANELLING AT G

SCALE OF INCHES FOR DETAILS & FEET FOR ELEVATION

ELEVATION

DETAIL OF ARCHITRAVE AT D

KENSINGTON PALACE: DETAILS OF DOOR IN KING'S GALLERY

PLATE XCV

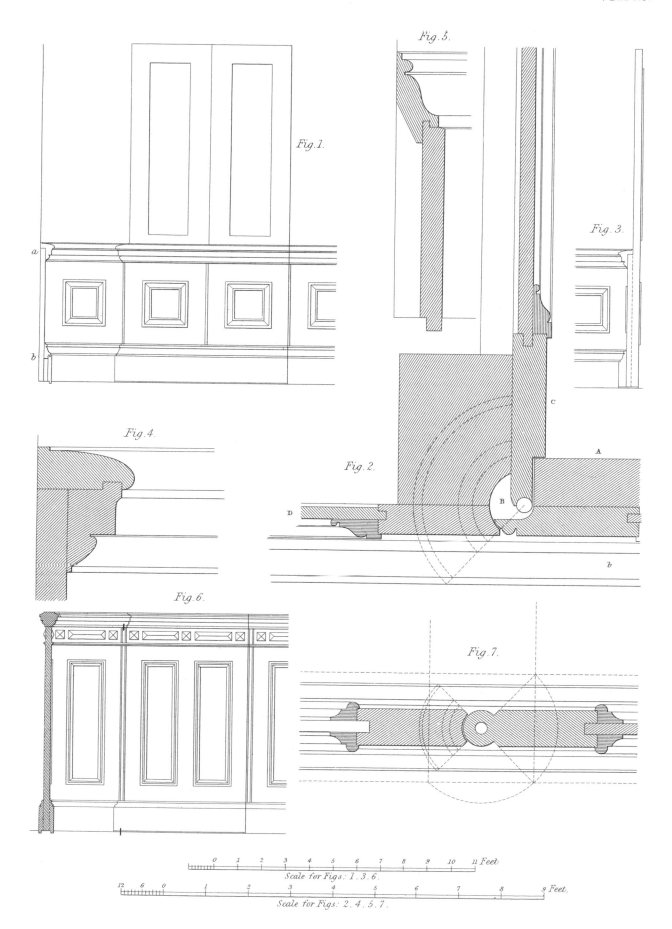

Fig.1.

Fig.2.

Fig.3.

Fig.4.

Fig.5.

Fig.6.

Fig.7.

Scale for Figs: 1.3.6.

Scale for Figs: 2.4.5.7.

JIB AND PEW DOORS

PLATE XCVI

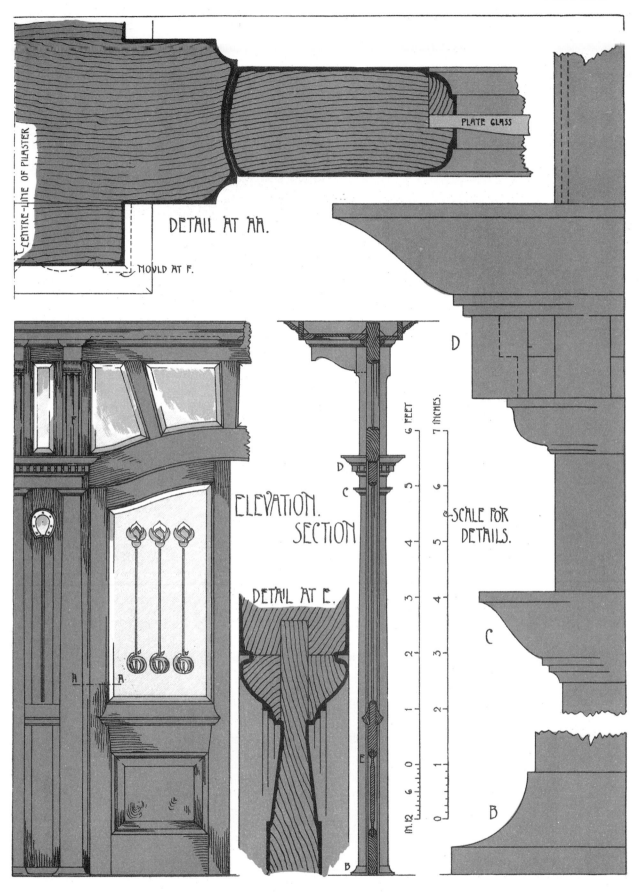

PLATE GLASS

CENTRE-LINE OF PILASTER

DETAIL AT AA.

MOULD AT F.

ELEVATION.
SECTION

DETAIL AT E.

D

C

B

6 FEET

7 INCHES.

SCALE FOR
DETAILS.

VESTIBULE DOORS AND SCREEN FOR HOTEL

$\frac{3}{4}$ inch thick; this ground ought to be wider than shown, in order to afford firmer support to the architrave. No. 3 is a pair of folding doors in which stop-chamfering is freely used; the enlarged horizontal section at B shows the general construction, the chief point of interest being the rebated meeting-stiles, beaded on one side and V-jointed on the other.

Two panelled and moulded doors are given in fig. 1121. In No. 1 the door has a segmental head, and the architraves at the sides of the door have three-quarter round shafts planted on (see detail at A) with moulded bases and caps. No. 2 is of more elaborate character in the Greek style; the details of construction are shown in the vertical section at B. The scales given in fig. 1120 apply also to this illustration.

Fig. 1122 is a French door of the fifteenth century, $1\frac{7}{8}$ inch thick and 3 feet $4\frac{1}{2}$ inches wide, and framed together in the usual way; but mouldings are planted on

Fig. 1121.—Two Panelled and Moulded Doors

No. 1 *No. 2*

the face, as shown in the detail at A. The panels are filled with tracery carved out of the solid.

An English four-panelled oak door of the seventeenth century is illustrated in fig. 1123. The framing is $1\frac{3}{8}$ inch thick, and the tenons of the rails do not go through the stiles, but are secured with oak pegs as shown. The panels are raised and carved in low relief. The framing is moulded on the solid, and the muntins are moulded on the face as well as on the angles. The furniture of the door is in bright iron of good design.

The example given in fig. 1124 is on similar lines, but more freely ornamented. It is a French door of the sixteenth century, and originally measured 2 feet $9\frac{1}{2}$ inches in width and 5 feet 8 inches in height. The thickness of the framing is only 1 inch, and the tenons do not go through the stiles, but are secured with pegs. The muntins, stiles, and rails are "square", but the muntins are square-sunk and fluted on the face (see detail at A), and the stiles and rails are square-sunk, the stiles being also carved. The panels are raised and moulded, and carved in low relief. The back of the door is similar to the face.

The door and doorway shown in Plate XCII furnish an interesting example of English work of the eighteenth century. Originally they formed the principal entrance to a house in

Lincoln's Inn Fields, London, but are now in the South Kensington Museum. The panels of the door are raised, and the chief peculiarity of the framing is in the diagonal rails of the bottom panel. The doorway is of elaborate design, richly carved and moulded.

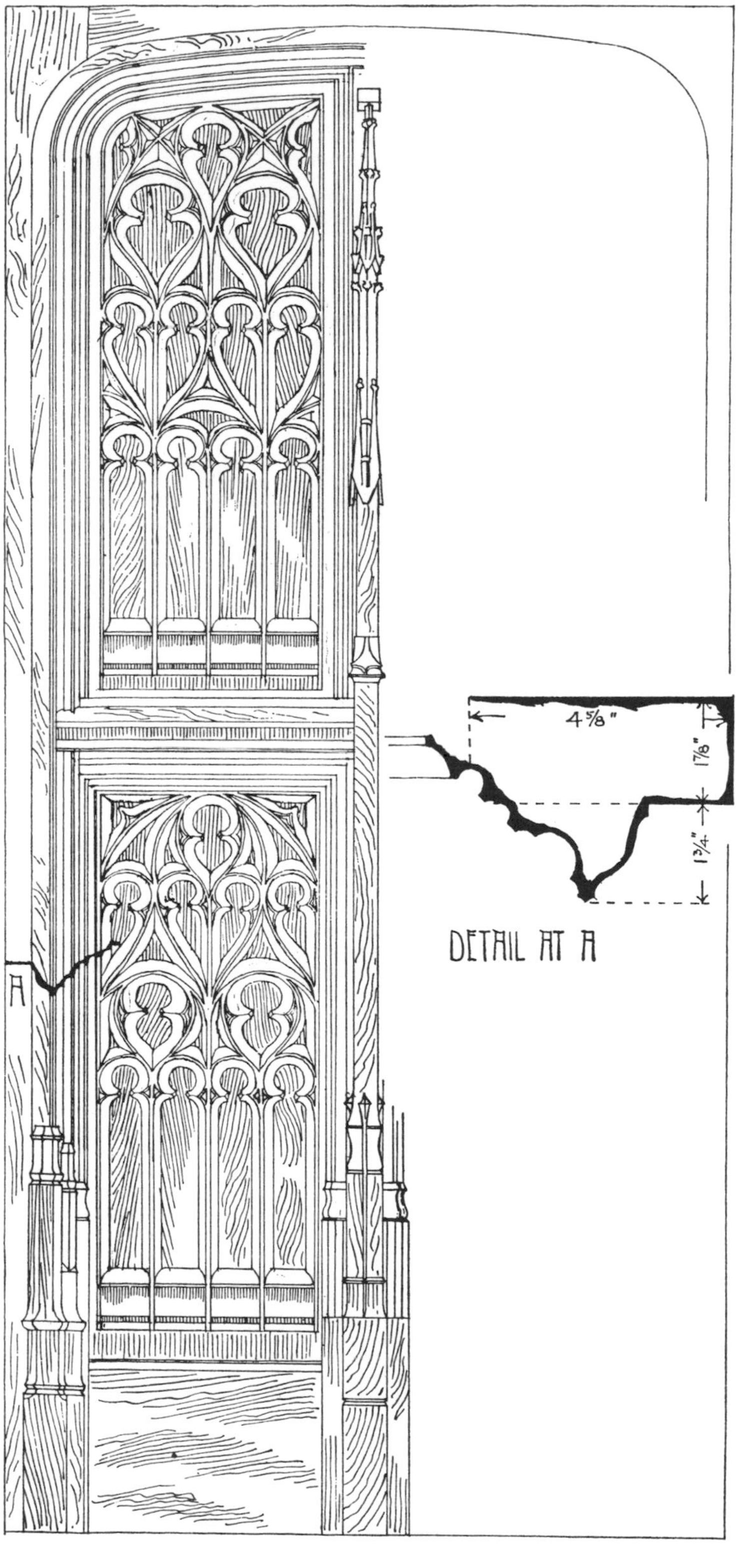

PLATE XCIII.—Fig. 1, No. 1, is the elevation of a double-margined door. No. 2 is an enlarged horizontal section on the line A B of No. 1. A A is the double-margined stile formed by the two stiles A A, *e e* the centre bead, B the panel, *c c* the mouldings laid in. The dotted lines show the manner in which the meeting-stiles are forked on the top and bottom rails, which is also represented in detail in fig. 1, No. 3, where B is an elevation of part of the top rail, thinned to enter the fork of the stiles, and C part of the stile in elevation, and A is the plan at the end of the stiles, showing the fork, the thinning of the rail, and the keys *a a*. Sometimes the double stile or muntin is in one piece, instead of two as shown, and the bead is run on the solid.

Fig. 2, No. 1, is a two-leaved or double-margined door, and fig. 2, No. 2, a section through the meeting-stiles A, B, showing the rebates *d d* and beads *e e*, which in this case are planted on. The panel D is let into a groove in the moulding C, in the French manner already noticed, and the moulding is framed and united as shown in fig. 2, Nos. 3 and 4, the latter being the end elevation of the angle at *d* (No. 3). A frieze rail should be introduced immediately below the level of the springing of wide semicircular-headed doors, as it is not an easy matter to make a perfectly safe joint between the curved top rail and the stile.

DETAIL AT A

4⅝"

1⅞"

1¾"

A

Fig. 1122.—French Door (late 15th Century)

Fig. 3, No. 1, is a double-margined Gothic door. Fig. 3, No. 2, is an enlarged section through the meeting-stiles showing the rebates A B, the panels D, and the panel-moulding C,

framed as in the last example. Fig. 3, No. 3, is an enlarged section through one of the panels, A stile, B centre rib, C moulding, D panel. As this method of forming the panels leads to considerable difficulty in the traceried heads, the bars are sometimes planted on, but the best way is to carve the tracery out of the solid, as in fig. 1122.

Where folding doors are required for narrow openings, the framing may with advantage be arranged as shown in fig. 1125. It will be seen that one leaf has a wider meeting-stile than the other, and that the difference between the two is concealed by the bead *a*; for

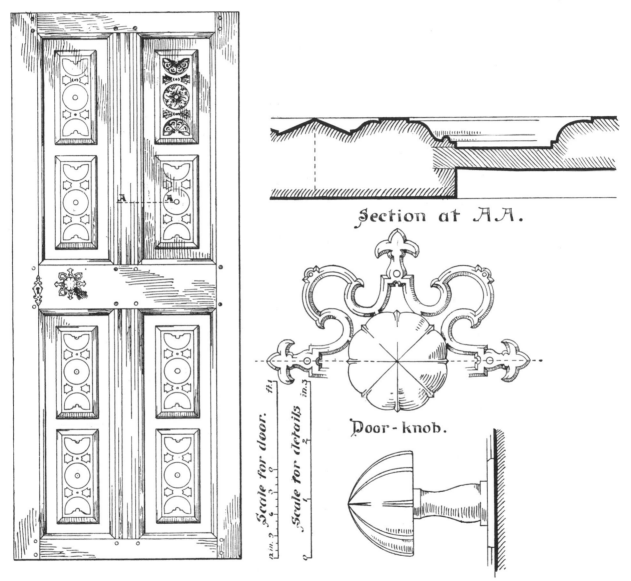

Fig. 1123.—English Door (17th Century)

ordinary purposes the wider leaf only is opened, the other leaf being fixed by bolts. By this arrangement the larger leaf will be, say, 4 inches wider than the other, and will thus give 2 inches more to the ordinary opening than if the usual arrangement had been adopted.

The double-margined door illustrated in Plate XCIV is an interesting example in the King's Gallery at Kensington Palace. The panels are raised and the panel-mouldings carved; the tenons of the rails are secured to the stiles by wood pegs. The architrave is moulded and carved, and is surmounted by a carved frieze, and carved and moulded cornice.

PLATE XCV.—Fig. 1 is the elevation of a jib door in the side of an apartment which has a base *b*, dado and dado moulding *a*; and fig. 3 is a vertical section through the door,

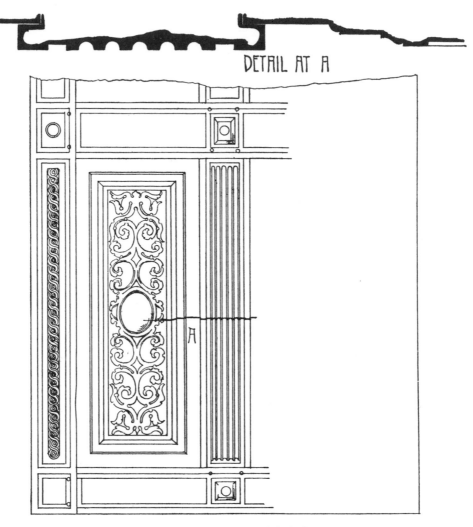

DETAIL AT A

Fig. 1124.—French Door (16th Century)

Fig. 1125.—Folding Doors with unequal Meeting-stiles

In fig. 2 is drawn to a larger scale a horizontal section through part of the hanging stile A of the jib door and the frame B in which it is hung. The panelled lining is shown at C. The dotted lines show the line of the hinging through the base mouldings b. Fig. 4 is a section through the dado moulding, and fig. 5 a section through the skirting and base moulding.

Fig. 6 is an elevation of a pew door, and fig. 7 a section through its hanging stile and the stile of the framing to which it is hung. It is treated precisely as a jib door.

A pair of folding doors and a screen for a vestibule, all in mahogany, are shown in Plate XCVI. The hotel for which they were designed is known as the White Horse Hotel, and the horse-shoes were introduced instead of a more conspicuous sign. The doors are of the swing type, with "Climax" spring hinges. The upper panels are glazed with embossed and bevelled plate-glass, which is bedded in wash-leather and secured with small wood mouldings in the manner shown in the details; the framing is moulded on the solid on both sides, except for the glazed panels, where the framing is moulded

on the solid on one side only, the other side being rebated to receive the glass and the loose moulding. As glazed panels are easily broken, the loose mouldings should be fixed with brass screws and countersunk cups, so that the mouldings can be taken out and refixed without injury. In making swing doors particular care should be taken that the edges of the meeting-stiles are perfectly true in line, and that they very nearly touch each other when the doors are closed; a wide and irregular joint shows bad workmanship. The edge of the hanging stile is rounded, and fits into a corresponding hollow in the frame. These curves should be struck with a radius equal to the distance from the spindle of the hinge to the heel of the door. The construction of the screen will be understood from the drawings.

Fig. 1126 shows a pair of panelled doors made to slide into recesses formed in the partition. No. 1 is the elevation, No. 2 the plan, No. 3 the vertical section, and No. 4 a section (to a larger scale) showing the top rail of the door, the recess in the partition, one pair of pulleys and the strap by which the door is hung to the pulleys, and the iron rails on

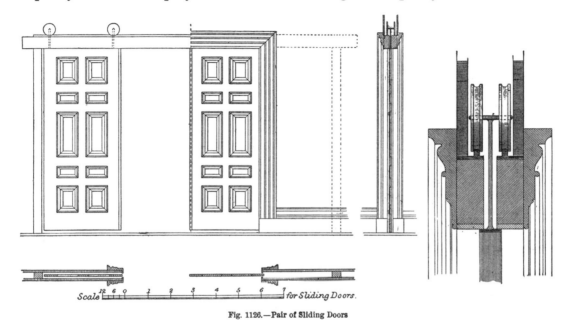

Fig. 1126.—Pair of Sliding Doors

which the pulleys travel. Other methods of hanging sliding doors will be illustrated in the last part of this chapter.

The ornamental doors which have hitherto been described are all of the framed type, but in many ancient and modern examples the doors are formed with plain boards, on which mouldings are planted to give the desired relief. The south door of Blythburgh Church, in Suffolk (fig. 1127), is an example dating from the fifteenth century. The door is formed with vertical boards 1 inch thick, nailed to $\frac{1}{2}$-inch horizontal boards behind. The construction is therefore of the same type as the ordinary ledged door, the difference being that the whole of the back of the door is covered with horizontal boards, instead of only three horizontal boards or "ledges" being used. On the face-boarding mouldings are planted to form the stiles, muntins, tracery, &c. These mouldings are in two thicknesses—the inner being $1\frac{1}{8}$ inch thick, and the outer $\frac{7}{8}$ inch. They are secured with oak pegs. The heading joints of these members (at the level of the springing) are splayed and secured with pegs. The doors are hung folding by means of wrought-iron strap hinges to hooks let into the stonework. The mouldings are planted on the door about $2\frac{1}{4}$ inches from the edge to form a rebate (as shown in the detail), which fits against the stonework of the doorway.

The same method of construction is seen in the south door of St. Helen's Church, Bishopsgate, London, shown in Plate XCVII. The doorway in which this occurs is dated

1633, and the door itself appears to be of the same age.   Each leaf of the door is formed with two vertical boards 1¾ inch thick (one much wider than the other), which are joined together edge to edge, probably by a loose tongue.   On the face of these main boards pieces ¾ inch thick are planted to form the stiles and rails, and on these again mouldings are planted in continuation of the mouldings of the stone capitals, &c.   The raised portions of the central panels are also planted on the main boards.   The pilaster on the meeting-stile is

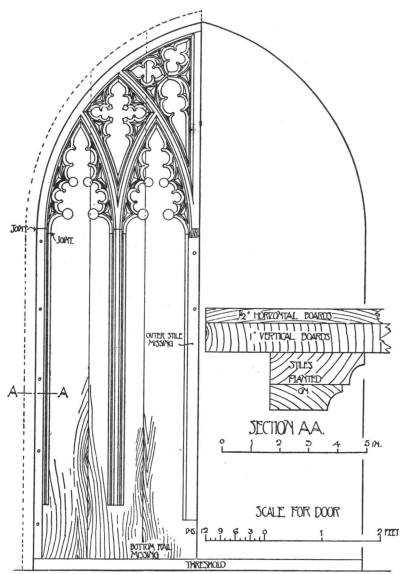

Fig. 1127.—South Door, Blythburgh Church, Suffolk

6¼ inches wide and ¾ inch thick up to the springing, and is ornamented with rolls, &c., planted on. In the head of the door the ornament is bolder; the curved mouldings with scroll have a greatest projection of 3½ inches, and the upper part of the pilaster projects 3 inches. The doors are of oak, and are hung with wrought-iron strap hinges to hooks fixed to the stonework.

Solid doors formed with two or three thicknesses of match-boards may be made with an outer sheathing of oak or teak, and an inner sheathing of deal or other wood, crossing each other at right angles. Sometimes the outer boards run diagonally instead of vertically.

Doors with glazed panels are often known as *sash-doors*. In many cases they are ordinary four-panel doors, but the framing around the upper panels is rebated to receive the glass, which is secured by a loose bead or moulding planted on and mitred round, as shown in Plate XCVI.   Other designs for glazed doors are given in Plates XX and XXI. Sometimes the stiles are narrowed in the upper part, in order to obtain a greater area of glass.   The difference in width is usually effected in the depth of the lock-rail, the end of which is splayed to fit the tapering part of the stile, as shown in No. 4, fig. 1030.   The glazed panel may be divided into squares or other figures by rebated sash-bars.

Fire-resisting doors are now often made with three thicknesses of boards, the inner being horizontal and the others vertical, nailed together and entirely covered with sheets of tinned steel (fig. 1153).   More recently such doors have been made with two thicknesses of boards, between which a sheet of " uralite " (prepared from asbestos) is placed; sheets of uralite are also nailed on both sides and around the edges of the woodwork, and

the whole is encased with steel sheets. A door of this kind, tested by the British Fire Prevention Committee, remained perfect on one side, although the side exposed to the fire was seriously damaged.

Revolving doors are now sometimes used, as shown in fig. 1128. In plan the doors form a cross with equal arms; they are pivoted at the top and bottom on the vertical axis by means of special fittings, so that the whole will revolve like a turnstile. The doors are

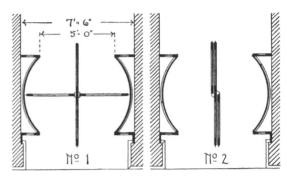

Fig. 1128.—Revolving Doors

enclosed at the sides by framing, each being in plan rather more than a quadrant of a circle. The advantages of these doors are that they are less noisy than swing doors, and also less draughty. In the Van Kannel revolving doors, the leaves are hinged so that they can be easily folded, as shown in No. 2, leaving two clear passages; in the standard size the vestibule is 7 feet in diameter and 7 feet 6 inches high inside. India-rubber tongues are inserted in the edges of the doors as shown, to prevent fingers being trapped and to exclude draughts.

Door-frames and Linings, &c.—Doors are hung to frames and linings of various kinds. For external doors the frames are usually formed from stuff not less than 3 inches thick or $4\frac{1}{2}$ inches wide, and rebated (A, fig. 1129) to a depth of $\frac{1}{2}$ to $\frac{3}{4}$ inch to receive the door; to afford additional protection against the weather, the rebate is often checked out as at B. The exposed edge of the frame at C may be square, chamfered, rounded, beaded, or moulded. The two jambs or legs of the frame are tenoned into the head and secured with wedges, and the head is often made to project 3 inches beyond the jamb, forming a "horn" which is built into the wall, as at D, No. 3. If a transome is introduced, it is tenoned and wedged

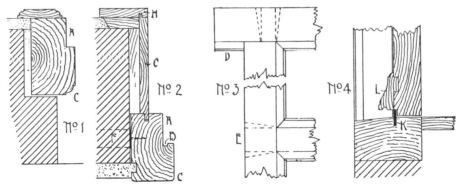

Fig. 1129.—Door-frames

to the jambs as at E. The glass in the fanlight above the transome may be fitted into the rebates of the frames and secured with beads or mouldings planted in, or a separate casement may be made and fitted into the rebates.

In some parts of the country it is customary to place the frames in position (one or more rough boards being first nailed across to keep them rigid) and to build the brickwork up to them. In such cases the frames are usually fixed by means of hoop-iron ties, one end of each being turned up or down and nailed to the frame, and the other part (12 or 15 inches long) being built into the brickwork; to protect them from rust, the ties should be dipped in tar, and to afford a better key for the mortar they should be sprinkled with sand before the tar has set. At the least three ties should be used for each jamb. In other parts of the country the frames are not inserted until the "carcase" of the building is finished, and this is the better method, at any rate in localities with a heavy rainfall. It was at one time

customary to build into the reveals of the openings a number of "wood bricks"—*i.e.* pieces of wood of the same dimensions as the building bricks used, or slightly larger—and to these the frames were nailed; but wood bricks absorb moisture from the mortar and consequently swell, and as the building gradually "dries out", they shrink and so work loose. Bricks or "fixing blocks" made of coke-breeze concrete are now preferred, as they can be nailed to and do not shrink; they should not be more than 18 inches apart vertically. Another method of fixing is to rake out some of the mortar-joints in the brick reveals, drive wedges between the bricks, and nail the frames to these wedges, but the wedges often work loose. The feet of the jambs may simply rest on the step or threshold, or may be fixed by dowels (one to each jamb and preferably of slate or copper) let into the step and projecting upwards into mortises in the jambs; or cast-iron sockets may be used, having the same external cross-section as the frame, in which case the foot of the frame must be sunk all round to a depth equal to the thickness of the metal, thus forming an irregular tenon which fits into the socket. Oak thresholds are sometimes used into which the frames are tenoned; holes should be bored from the face of the threshold to the bottom of the mortise to allow moisture to escape. To prevent rain driving under the door, a wrought-iron bar is sometimes fixed to the threshold, as shown in No. 4, fig. 1129, at K; a weather-board L, throated underneath, may also be used. Occasionally a stone is built into the wall above the step, the portion projecting beyond the reveal being rebated and moulded exactly like the frame; the foot of the frame is thus raised above the step and decay is less likely to occur.

As a rule, external door-frames are fixed at some distance back from the face of the wall, as shown in No. 1, fig. 1129, rebates being formed in the brickwork to receive the frames. No. 2 shows a frame suitable for a rough-cast wall, a groove being run in the frame to afford a key for the rough-cast and to make a weather-tight joint; this frame is also grooved on the inner side to receive the linings at G; the architrave is shown at H, and is fixed with nails to the linings and to the rough ground, which is splayed on the edge to form a key for the plaster. If the reveal on the inner side of the frame is of sufficient depth, the linings may be framed and panelled.

Solid frames of this kind are sometimes used for heavy internal doors, but as a rule internal doors are hung to linings or "casings", which seldom exceed $1\frac{7}{8}$ inch in thickness,

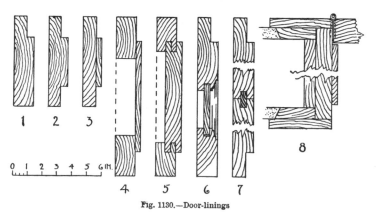

Fig. 1130.—Door-linings

and are usually only $1\frac{3}{8}$ inch. For walls not exceeding 9 inches in thickness, the linings may be formed from single widths of stuff, those for $4\frac{1}{2}$-inch walls (plastered on both sides) being 6 inches wide, and those for 9-inch walls $10\frac{1}{2}$ inches wide. The linings may be *single-rebated*, as in No. 1, fig. 1130, or *double-rebated*, as in No. 2 (the second rebate being formed for the sake of appearance), or "stops" may be planted on to form the rebate or rebates, as in No. 3. Nos. 4 and 5 are modifications known as "skeleton linings", and are suitable for 14-inch walls. A panelled lining is shown in No. 6. If plain deal linings are used in walls more than 9 inches thick, they must be cross-tongued as in No. 7; warping may be prevented by means of hardwood keys (slightly tapered and with bevelled edges) driven into dovetailed grooves across the back of the lining. The jambs are housed or tenoned into the head, and nailed to wood bricks, fixing blocks, or wedges in the same way as external door-frames, or to rough grounds. If the door must be flush with the face of the architrave, the inner edge of the latter will form part of the rebate,

PLATE XCVII

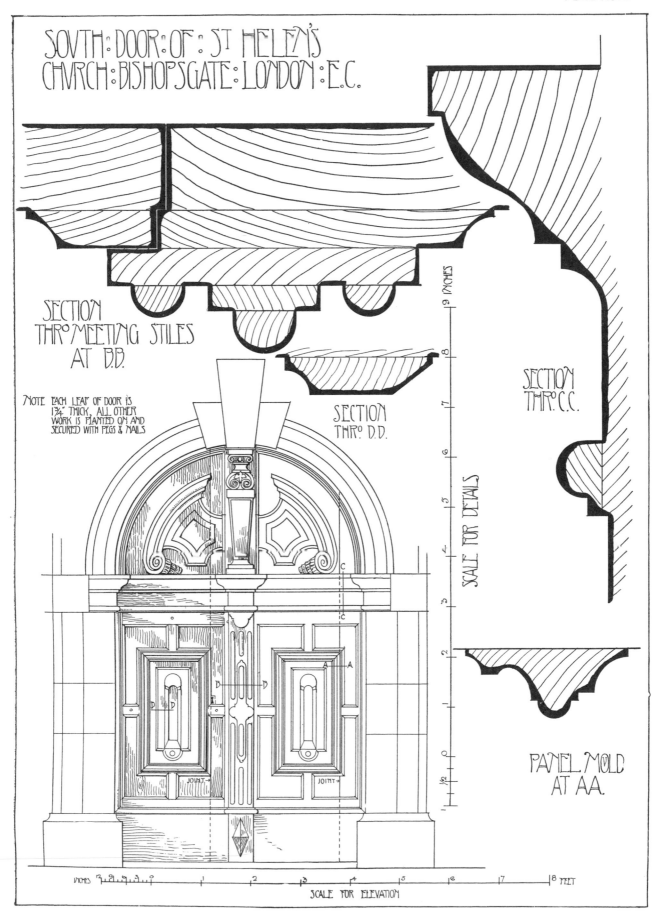

SOVTH·DOOR·OF·ST HELEN'S
CHVRCH·BISHOPSGATE·LONDON·E.C.

SECTION
THRO·MEETING STILES
AT B.B.

NOTE EACH LEAF OF DOOR IS
1¾" THICK, ALL OTHER
WORK IS PLANTED ON AND
SECURED WITH PEGS & NAILS

SECTION
THRO· D.D.

SECTION
THRO·C.C.

SCALE FOR DETAILS

9 INCHES

PANEL MOLD
AT A.A.

JOINT   JOINT

INCHES   1   2   3   4   5   6   7   8 FEET
SCALE FOR ELEVATION

SOUTH DOOR OF ST. HELEN'S CHURCH, BISHOPSGATE, LONDON

as shown in No. 8; this arrangement is suitable for doors intended to open back against the wall, but care must be taken that the screw-holes of the butt-hinges do not come on the joint between the architrave and the lining.

Rough grounds of the same thickness as the plaster (usually $\frac{3}{4}$ inch) should be fixed around all internal doorways in plastered walls to afford a proper fixing for the linings and architraves, and to serve as " screeds " or guides for floating the plaster to. The width of the ground must be regulated by that of the architrave, as the latter ought to cover the joint between the ground and the plaster. The outer edge of the ground is usually splayed, as in No. 8, fig. 1130, to give a key to the plaster. The angles are often framed together, one method known as *bevelled haunching* being shown at A, fig. 1131. In thick walls, *backings* are often fixed between the grounds at intervals, as at B, to which the linings can be nailed. The grounds are nailed to wood bricks, fixing blocks, or wedges in the manner already described.

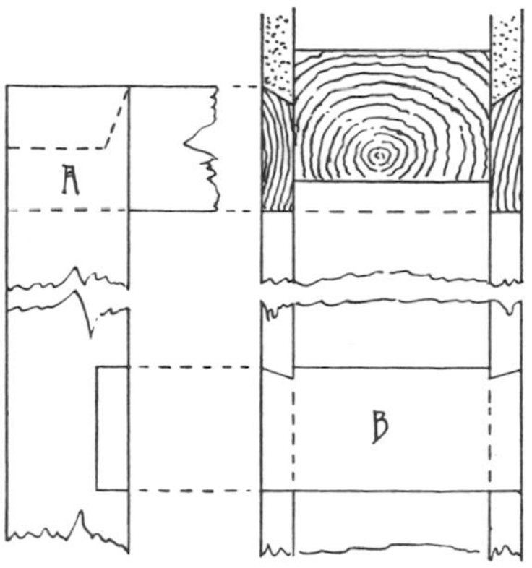

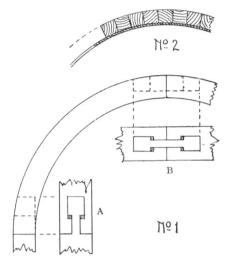

Fig. 1132.—Arched Door-frame and Lining

Arched heads in door-frames may be formed as shown in No. 1, fig. 1132, the joint at the springing being formed by means of a necked tenon A, drawn tight by wedges, and that at the crown by a hard-wood key B, also secured with wedges. A handrail bolt is sometimes used instead of the hardwood key,

Fig. 1131.—Framed Grounds with Backings

and is more easily fixed, but the joint is not as strong. Arched linings may be formed from boards of the same thickness as the jamb-linings, by making transverse saw-cuts on the face to a sufficient depth to allow the boards to be bent to the required curve, but this method is not satisfactory in any case, and is altogether unsuitable for hardwoods. A better method consists in using a face-board of the wood specified, thin enough to be bent around a template of the required curve; it is temporarily secured in this position, and the back is then covered with transverse strips of wood glued to the face-board and to each other, as shown in No. 2, fig. 1132; these strips form a fairly rigid backing when the glue has set, but in fixing the lining, wedges should be inserted between it and the brickwork to prevent any alteration of the curve.

Architraves.—Doorways are in general surrounded by an ornamental wooden margin or border, not merely, however, for ornament, but especially to cover the junction between the plaster and woodwork. These margins are sometimes plates of wood, plain, chamfered, or ornamented with mouldings, and are termed architraves, examples of which are given in fig. 1105 and Plate XCVI, &c., or they may consist of pillars or pilasters with proper entablature, as in Plate XCIV. The pilasters are usually set on solid blocks of the same height as the skirting, and so also are the architraves in good work; but in other cases they run down and are scribed to the floor. Sometimes the outer edge of the architrave is moulded in the same manner as the skirting, and the two are then mitred together (fig. 1174, No. 3). Architraves ought to be fixed to rough grounds as already described. The ground being fixed and the door-frame set in its place, serve as guides for floating the plaster by, and when the plaster is dry the architrave should be applied so as to lap over the joint and effectually cover it; it is a common error to fix the architraves before the plastering is complete.

Folding Partitions are constructed in the same way as doors, and differ principally in

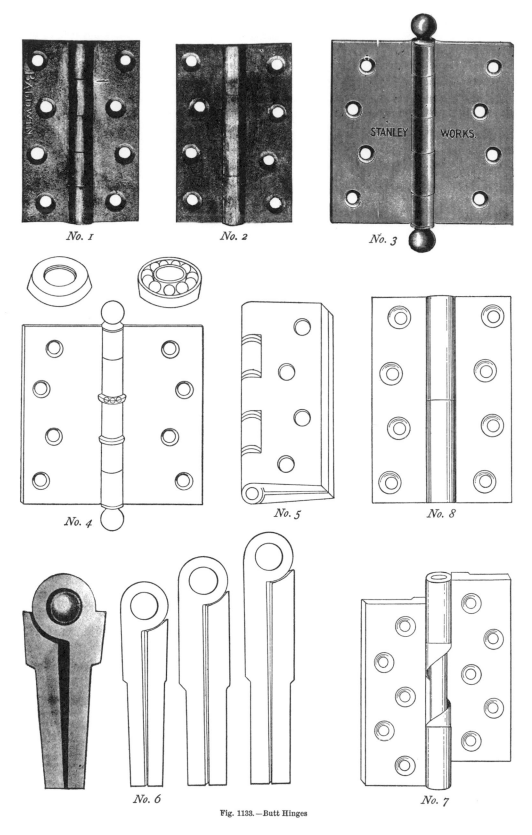

No. 1   No. 2   No. 3

No. 4   No. 5   No. 8

No. 6   No. 7

Fig. 1133.—Butt Hinges

the fittings required.   They may be hung from the top like sliding doors, or may be
supported at the bottom on swivel-pulleys running on iron tracks in the floor; the latter
method is now generally preferred, as there is less strain on the framing.   Plate XCVIII
shows a partition of this kind designed by the writer for a school.   The wood throughout

PLATE XCVIII

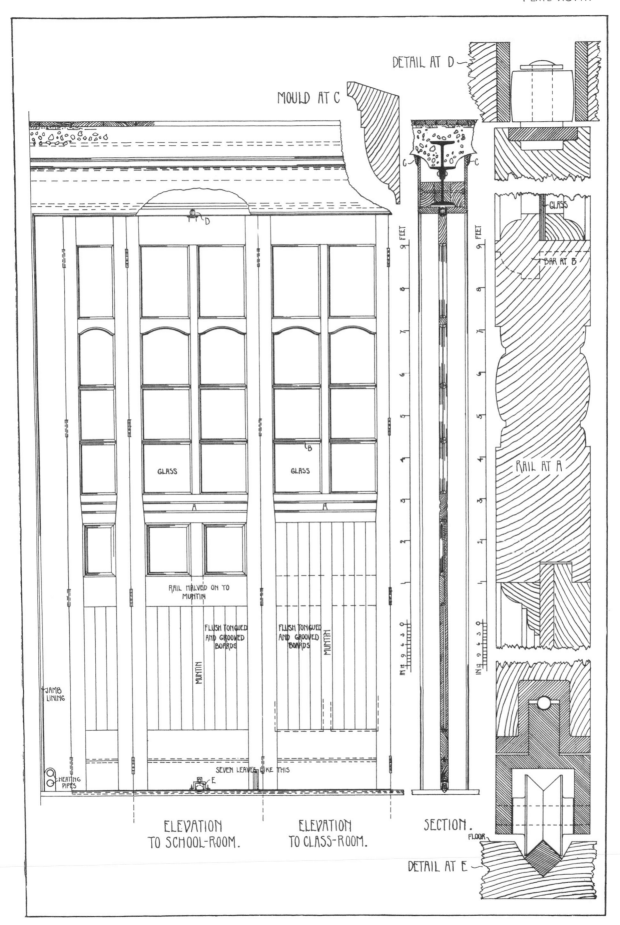

DETAIL AT D

MOULD AT C

DETAIL AT E

GLASS

BAR AT B

RAIL AT A

FLOOR

ELEVATION
TO SCHOOL-ROOM.

ELEVATION
TO CLASS-ROOM.

SECTION.

GLASS

GLASS

RAIL HALVED ON TO
MUNTIN

FLUSH TONGUED
AND GROOVED
BOARDS

FLUSH TONGUED
AND GROOVED
BOARDS

MUNTIN

MUNTIN

JAMB
LINING

HEATING
PIPES

SEVEN LEAVES LIKE THIS

FOLDING PARTITION IN SCHOOL

is of yellow pine, and the framing is 2 inches thick. The lower portions are flush panelled on both sides so that the partition can be used as a blackboard.

Hinging.—Hinging in joinery is the art or operation of attaching two pieces of wood together, such as a door to its frame, a shutter to the window-lining, or a back flap to a shutter, by metal fastenings that permit one of the pieces to revolve. The fastenings are known as hinges.

No. 1

Fig. 1134.—Parliament and Pew Hinges   *No. 2*

The form of hinge most frequently used is the *butt*. It is made of cast-iron, wrought-iron, steel, or brass in a variety of forms and sizes. Butts are classified and sold according to the height; thus, a 3-inch butt is 3 inches high. The width (measured across the hinge when open) increases with the height, but hinges of greater width than the ordinary butts are made and sold as "broad butts"; in these the width usually equals the height, but in some cases exceeds it. For light internal doors (say 2 feet 6 inches by 6 feet 6 inches by $1\frac{1}{2}$ inch) one pair of 3-inch butts will suffice, while for larger doors 4-inch or 5-inch butts must be used. Heavy and lofty doors often have $1\frac{1}{2}$ pair of 5-inch butts.

No. 1

No. 2

Fig. 1135.—Back-flap and Counter-flap Hinges

No. 3

Cast-iron butts (No. 1, fig. 1133) are by some makers classified as "light", "medium", and "strong", the first being made in the following sizes,—1, $1\frac{1}{4}$, $1\frac{1}{2}$, $1\frac{3}{4}$, 2, $2\frac{1}{4}$, $2\frac{1}{2}$, 3, $3\frac{1}{2}$, 4, and 5 inches; "medium"—2, $2\frac{1}{4}$, $2\frac{1}{2}$, 3, $3\frac{1}{2}$, 4, $4\frac{1}{2}$, 5, and 6 inches; and "strong" in the same sizes as "medium" except $2\frac{1}{4}$ inches. The joints are usually polished in order to reduce friction.

Wrought-iron butts are more expensive than cast, and are made with single joints (No. 2, fig. 1133), or double joints similar to the cast-iron butt shown in No. 1.

Pressed-steel butts of ordinary strength are much cheaper than cast-iron, but "heavy" butts with polished joints are more costly than cast-iron butts of the same sizes. Pressed-steel butts range from $1\frac{1}{2}$ inch to 4 inches, increasing by $\frac{1}{2}$ inch; the 5-inch size is also made

Fig. 1136.—Centre Hinges

Wrought-steel butts are still more expensive, the sizes increasing by $\frac{1}{2}$ inch from 3 to 5 inches; the 6-inch hinge is also made in certain forms. No. 3, fig. 1133, shows a wrought-steel butt with loose pins, instead of the usual fast joint; the pins are screwed into the top and bottom knuckles of one leaf of the hinge. No. 4 shows a wrought-steel butt with washers and ball-bearings; this form of hinge is particularly suitable for heavy doors, as the ball-bearings ensure easy action.

Brass butts may be either "machine-made" or "cast", the former being made in sizes from $\frac{1}{2}$ inch to 3 inches, and in some cases 4 inches, and the latter in sizes increasing by $\frac{1}{4}$ inch from 1 inch to 3 inches, and by $\frac{1}{2}$ inch from 3 inches to 5 inches; 6-inch cast-brass butts are also made. The widths are described as "narrow", "medium", and "broad"; heavier butts are known as "strong". The pins may be of iron or brass. In design ordinary brass butts do not differ from iron butts, but many are made with steel washers between the joints, as shown in No. 5, fig. 1133, as steel does not wear away as rapidly as brass.

Projecting butts (No. 6) are made in iron, steel, and brass, in sizes from 2 inches to 6 inches. Different widths can be obtained in each size; thus, 4-inch brass projecting butts are stocked in 3, $3\frac{1}{2}$, 4, $4\frac{1}{2}$, and 5-inch widths, the projection (measured from the face of the door to the centre of the pin) ranging from $\frac{1}{2}$ inch to $1\frac{3}{8}$.

Rising butts are a variety of projecting butts with a single joint of helical shape (No. 7), so that the door rises on being opened and tends to shut when released. In brass hinges

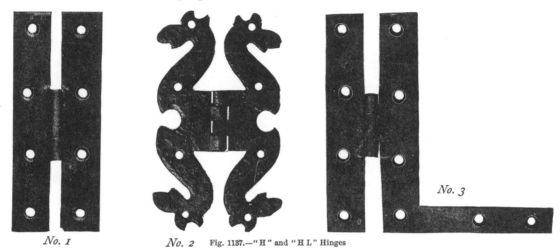

No. 1   No. 2   Fig. 1137.—"H" and "H L" Hinges   No. 3

the helical joints are faced with steel. In "skew rising" butts, the joint is **V**-shaped, formed with two helices. Another variety of rising butt has the pin in the form of a screw.

Loose or *lifting butts* (No. 8) have a single joint, so that the door can be lifted off without unscrewing the hinge, and are made of cast-iron or brass.

Hollow-joint butts can be obtained in brass, gun-metal, &c., for special purposes, such as doors with rebated edges, &c.

Parliament or *shutter hinges* are projecting hinges suitable for the external shutters of windows set in reveals; the projecting hinges allow the shutters when open to lie along the face of the wall. They are made in cast and wrought iron and brass. No. 1, fig. 1134, shows a cast-iron hinge of this type, made in three sizes, namely with 2-inch joint to open 4 inches, $2\frac{1}{4}$-inch joint to open $4\frac{1}{2}$ inches, and $2\frac{1}{2}$-inch joint to open 5 inches. Brass hinges are made to open from $2\frac{1}{2}$ to 6 inches.

Pew or Egg-joint hinges (No. 2, fig. 1134) are intended for small doors with projecting moulds, the projecting knuckle of the hinges allowing the doors to open back against the framing to which they are hung. They are made of brass, in sizes from 2 to $2\frac{1}{2}$ inches.

Back-flap and counter-flap hinges are shown in fig. 1135. No. 1 is a brass back-flap hinge, which is made in sizes from $\frac{3}{4}$ to 3 inches, and in "ordinary", "strong", and "extra strong" qualities. Nos. 2 and 3 are two forms of the brass counter-flap hinge, which is designed so that the flap when opened will lie flat on the counter; it will be seen that the hinge has two pins, each of which passes through the knuckle of one leaf of the hinge and through a short connecting link or "tumbler". Somewhat similar hinges for trap-doors are made of wrought-iron.

Centre hinges (fig. 1136) are sometimes used for the doors of pews, cupboards, and wardrobes. They are made of brass in sizes from 2 to 3½ inches.

"*H*" and "*HL*" *hinges* are shown in fig. 1137. These are screwed to the face of the work. The "H" hinge (No. 1) is of plain wrought iron, in sizes from 3 to 8 inches; No. 2 is one of the many ornamental varieties in wrought or malleable iron; No. 3 is a plain wrought "HL" hinge, and ornamental varieties of this type can also be obtained.

Cross-garnet or "T" *hinges* are of various forms, and are made of iron, pressed or wrought. No. 1, fig. 1138, shows one variety of pressed hinge; No. 2 is the "London" pattern; No. 3 the "Scotch"; No. 4 is a wrought-iron hinge; and No. 5 a variety known

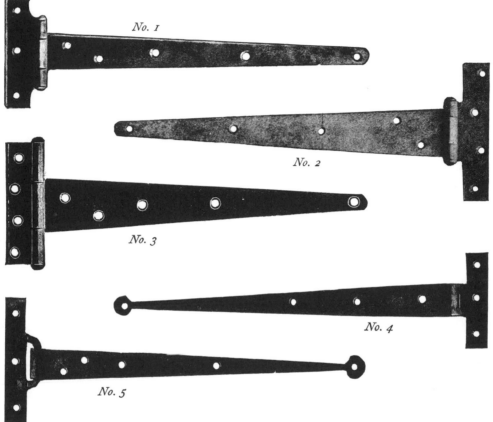

Fig. 1138.—Cross-garnet or "T" Hinges

as the "water-joint". Nearly all are made in sizes increasing by 2 inches from 6 to 20 inches; the 24-inch hinge is also stocked, and the No. 1 variety is made 30 inches long. Nos. 1 to 4 are generally used for ledged doors, and No. 5 for trap-doors.

Coach-house hinges (fig. 1139) are used for heavy framed doors and gates. No. 1 is the ordinary type of wrought-iron coach-house hinge, and No. 2 the *double-strap* hinge; the "hooks" on which the ends of the straps turn may have a back-plate as shown for screwing to a wood frame or post, or may be made with lugs for building into brickwork or with a dovetailed rag-bolt for letting into stone, the bolt being fixed with neat Portland cement or with lead. They are generally sold by weight. No. 3 is Collinge's patent gate or coach-house hinge, made of cast (malleable) iron, wrought iron, or steel, and in sizes rising by 3 inches from 1 foot 6 inches to 5 feet, and by 6 inches from 5 feet to 7 feet 6 inches. The step or hook of this hinge is cup-shaped, and in the cup a ball-shaped pin (attached to the end of the strap) works, the joint being covered by an inverted cup; the cup of the hook forms a receptacle for oil, and dust is excluded by a leather washer.

The hinges which have been described are those commonly used by the joiner, but

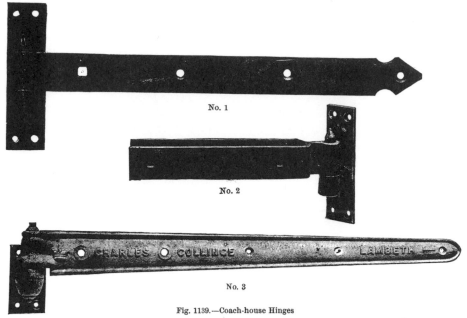

No. 1

No. 2

No. 3

Fig. 1139.—Coach-house Hinges

other varieties are made, and special hinges can be obtained to order. The methods of fixing remain to be considered.

Some examples of the use of butt hinges have been given in figs. 1043 to 1045. In fig. 1140 the hinges are fixed slightly askew; Nos. 1 and 2 show a door closed and open, and Nos. 3 and 4 a shutter for a window. This arrangement is adopted when the knuckle of the hinge is intended to run in line with a bead on the door or frame. Examples of this are shown in detail in fig. 1141; in I, the bead is on the frame A, and an air-tight joint is formed at a; in II, two beads are formed on the door—one on each side—and the hinge is set square; in III, one bead is formed on the door in line with the knuckle of the hinge, and another bead on the inside of the frame, the hinge being skewed; in IV, the hinge is square, and the beads are not opposite to each other. In each case, No. 1 shows the door closed, No. 2 shows it open, and No. 3 the jointing by the door and frame. Examples II, III, and IV are suitable for cupboard doors, being in frames of the same thickness.

Shutters for windows are a kind of doors, and are framed in a similar manner, and it will be convenient to consider the details of the hinging in this chapter. Fig. 1142 shows the ordinary method of hinging the shutter to the sash-frame; in No. 1 the shutter is turned back against the lining; in No. 2 it is turned in front of the window. Three methods of hinging a back-flap to a shutter are given in fig. 1143; in No. 2 the flap is (when closed) thrown

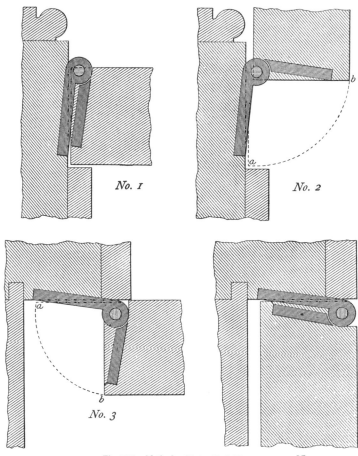

No. 1

No. 2

No. 3

No. 4

Fig. 1140.—Methods of fixing Butt Hinges

back from the joint, as shown by the dotted lines. The rule-joint, shown in No. 3, is used for the back-flap of a shutter which folds on to the face of the wall and not into a shutter-box, and also for the leaves of tables, &c.

Fig. 1144 shows the manner of finding the rebate when the hinge is placed on the

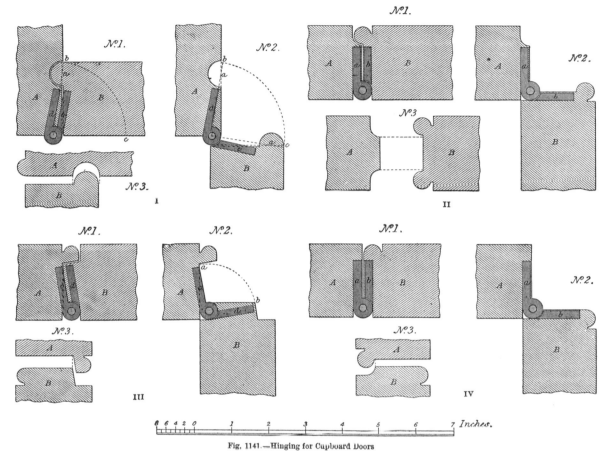

Fig, 1141.—Hinging for Cupboard Doors

contrary side. Let f be the centre of the hinge, ab the line of joint on the same side, hc the line of joint on the opposite side, and bc the total depth of the rebate. Bisect bc in e and join ef; on ef describe a semicircle cutting ab in g, and through g and e draw gh cutting dc in h, and join dh, hg, and ga to form the joint.

Fig. 1145 shows the method of fixing a projecting butt or pew hinge. In fig. 1146 the

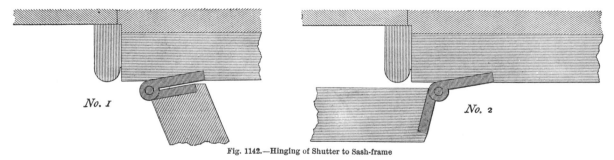

Fig. 1142.—Hinging of Shutter to Sash-frame

use of hollow-joint butts for a door to open to an angle of 90° is shown; it will be seen that the hinge is concealed when the door is closed. The segments are described from the centre of the pin g, and the dark-shaded portion requires to be cut out at each hinge to allow the two leaves of the hinge to meet.

Different methods of using centre-pin hinges are shown in Plate XCIX. Fig. 1 has the

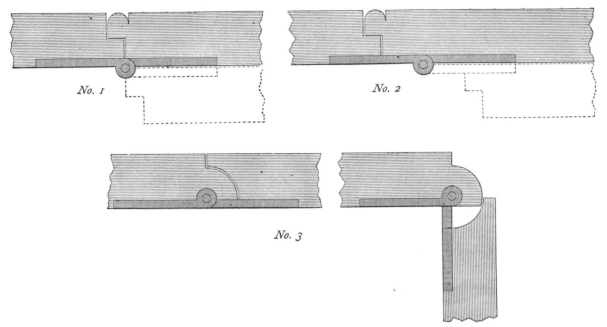

Fig. 1143.—Hinging of Back-flaps to Shutters

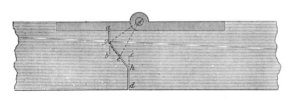

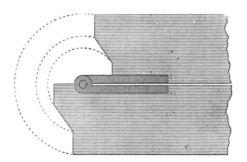

Fig. 1144.—Hinging of Back-flap with Rebated Joint

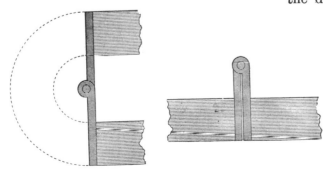

Fig. 1145.—Projecting Butt or Pew Hinge

edge of the door formed into a large bead or roll, which fits into a corresponding hollow in the frame; in this and the other figures on this plate, No. 1 shows the door closed and No. 2 open. In fig. 2 the frame is slightly hollowed, and a stop is planted on one side of the hollow to conceal the joint when the door is closed. In fig. 3 the frame is designed so that the edge of the door is concealed in whatever position the door may be. The same result is obtained in fig. 4, by means of a hinge having a strap B on the face of the door. Fig. 5 shows a centre-pin hinge for a swing-door, which allows the door to open either way, and to fold back against the wall in either direction. Draw ab at right angles to the door, and just clearing the line of the wall, or rather representing the plane in which the inner face of the door will lie when folded back against the wall; bisect it in f, and draw fd the perpendicular to ab, which make equal to af or fb, and d is the place of the centre of the hinge. Fig. 6 is another variety of centre-pin hinging opening to 90°. The distance of b from ac is equal to half of ac. In this, as in the former case, there is a space between the door and the wall when the former is folded back. In the succeeding figures this is obviated.

Fig. 7, No. 1. From a draw ab making an angle of 45° with ad; draw de and make eg equal to one and a half times ad; draw fg at right angles to ed, and bisect the angle

PLATE XCIX

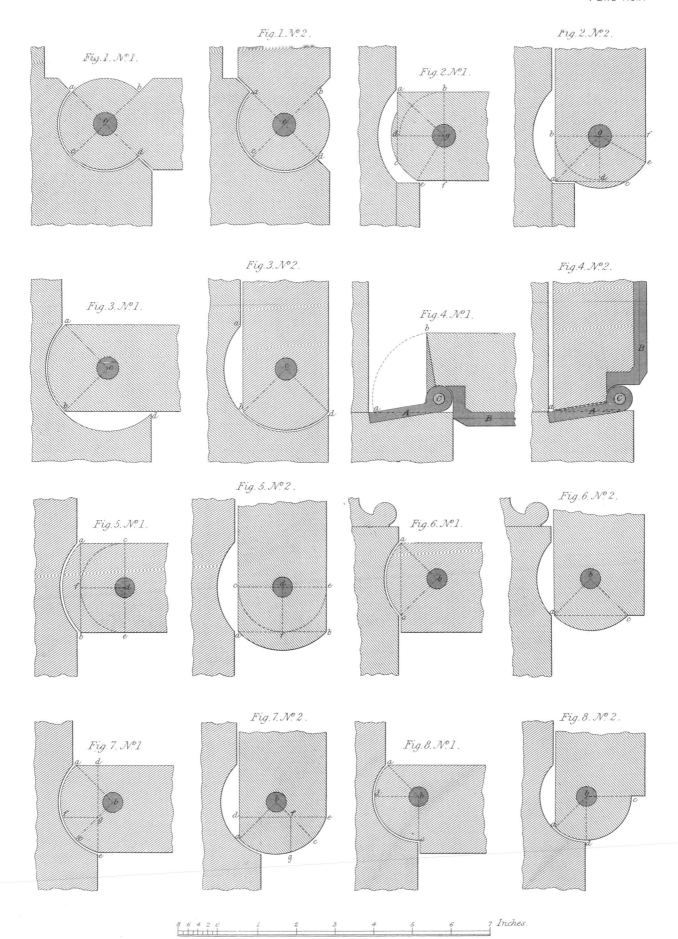

HINGING

f g e by the line *c g*, meeting *a b* in *b*, which is the centre of the hinge. No. 2 shows the door folded back, when the point *e* falls on the continuation of the line *f g* in No. 1.

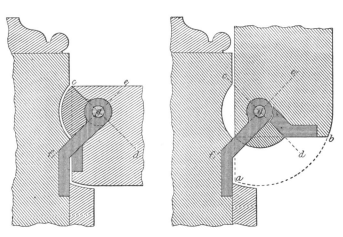

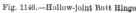

Fig. 1146.—Hollow-joint Butt Hinge

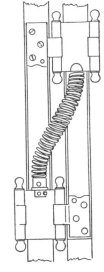

Fig. 1147.—Double-action Spring Butt

Fig. 8, Nos. 1 and 2. To find the centre draw *a b*, making an angle of 45° with the back of the door, and draw *c b* parallel to the jamb, meeting it in *b*, which is the centre of the hinge. The door revolves to the extent of the quadrant *d c*.

Spring Hinges.—For doors which are required to close automatically, some kind of spring hinge is generally used. This may be of the butt type, as in fig. 1147, which shows a double-action spring butt; this is made in sizes to fit doors from $1\frac{1}{8}$ to $3\frac{1}{2}$ inches thick, and is suitable for swing-through doors. Single-action butts of the same kind are made for doors opening to 90°. In

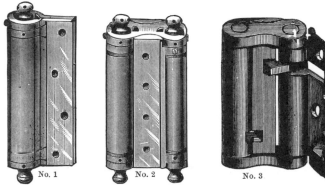

Fig. 1148.—Regulating Helical Spring Butts: No. 1, Single-action; No. 2, double-action; No. 3, blank

hanging a door, only one spring hinge is used, the other hinge being a "blank butt", the spring and blank butts being sold together

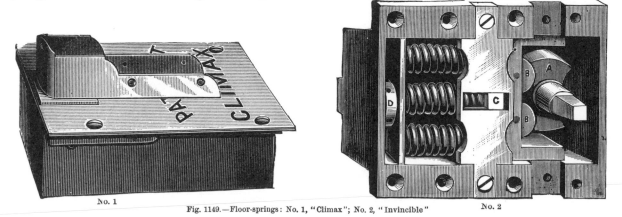

Fig. 1149.—Floor-springs: No. 1, "Climax"; No. 2, "Invincible"

as a pair. Fig. 1148 shows regulating helical spring butts, No. 1 being single-action, No. 2 double-action, and No. 3 a "blank". A more expensive but more satisfactory

hinge for the purpose is the floor-spring, which is made in endless variety for single and double action. A view of the "Climax" floor-spring is shown in fig. 1149, No. 1; a brass shoe is provided to receive the heel of the door, the shoe being attached to a spindle which passes down into the box in which the springs are placed. No. 2 shows the internal details of the "Invincible" floor-spring. The cam A is attached to the spindle which passes up through the top-plate of the box to the shoe in which the door is fixed. On the door being opened, the slide C, to which the wheels B B are attached, is pushed back against the springs, and on the door being released the springs force the slide back to its original position. D is a regulator for increasing or reducing the tension of the springs. The box containing the mechanism

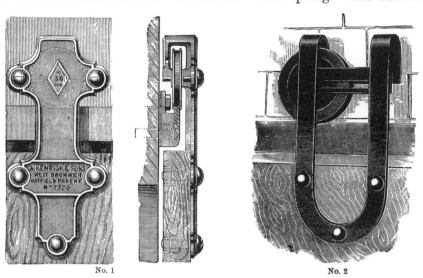

No. 1   No. 2
Fig. 1150.—Hatfield's Top Rollers for Sliding-doors

should be filled with good lubricating oil, so that the parts will work easily without much wear and tear. The top of the door is hung with a centre hinge. Some floor-springs are made with pneumatic, hydraulic, or oil checks to prevent slamming. The shoes and top centres are in some cases adjustable. Into the details of springs and other fittings fixed to the faces or backs of doors, we cannot enter.

Rollers, &c., for Sliding-doors.—An ordinary type of pulley and rail for sliding-doors has been given in fig. 1126. Hatfield's fittings for the same purpose work more easily; for top-hung doors, the fittings may be of malleable iron (No. 1, fig. 1150) or of wrought-iron (No. 2). The former are made in sizes with pulleys from 2 to 8 inches in diameter for runs from 2 to 12 feet, the rails increasing from $\frac{1}{4}$ by $1\frac{1}{2}$ inch to $\frac{5}{8}$ by 4 inches; the wrought-iron fittings have pulleys from 2 to 10 inches in diameter, runs from 2 to 20 feet, and rails from $\frac{1}{4}$ by $1\frac{1}{2}$ inch to $\frac{5}{8}$ by 4 inches. The special feature of these fittings is the slides or grooves in which the axles of the pulleys run. The same device is adopted in the rollers for the bottoms of sliding-doors, as shown in fig. 1151, where No. 1 is a fitting for mortising into the bottom rail, No. 2 a fitting for fixing into a recess in the side of the bottom rail, and No. 3 two forms of rail on which the pulleys run. No. 1 is made with pulleys from $1\frac{1}{2}$ to 8 inches in diameter, the length of run ranging from 2 to 20 feet; and No. 2 with pulleys from 2 to 10 inches, and runs from $2\frac{1}{2}$ to 40 feet.

No. 1

No. 2

No. 3

Fig. 1151.—Hatfield's Bottom Rollers for Sliding-doors

The "Coburn" trolley track and hangers, made in America, possess some valuable features. The track is of steel, the ordinary section being shown in fig. 1152, but double and treble tracks are made for two or three doors sliding side by side. The trolleys or

pulleys are of hard fibre or of iron, and are arranged in pairs, which run in the two channels of the track. In some varieties the pulleys are made with ball-bearings, and in others with roller-bearings. Fig. 1152 shows the fittings for a door sliding outside a wall; the strap

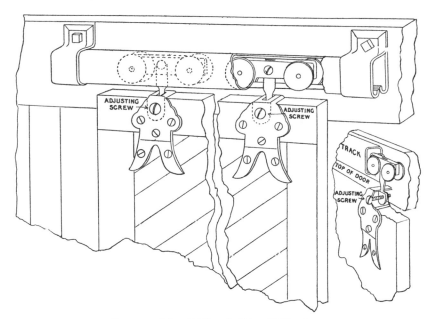

Fig. 1152.—"Coburn" Fittings for External Sliding-door

attached to the door is fitted with a screw for lateral adjustment, so that the door can be brought exactly under the track.

Fig. 1153 shows the application of somewhat similar hangers to a self-closing fire-resisting door constructed in the manner described on page 330. The door must be large

Fig. 1153.—"Coburn" Fittings for Gravity-closing Fire-resisting Door

enough to overlap the opening at least 3 inches at the sides and top, and the top rail must slope about $\frac{3}{4}$ inch to the foot. The track is fixed to the wall to the same slope. The fittings include two wall-stops A and D, a sheaf-plate C, lever and weight E, guide-roll F, floor-stop G, fusible link B with hooks and connecting cords.

Fig. 1154 contains vertical sections showing two of the arrangements adopted for a

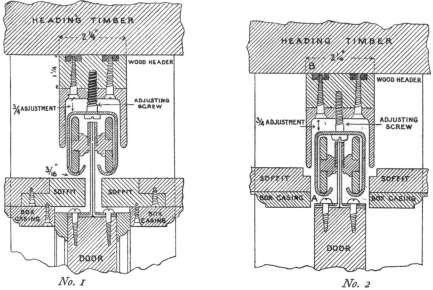

Fig. 1154.—"Coburn" Fittings for Door sliding into Pocket

single door sliding into a pocket or recess in a partition. A projecting stop is fixed on one edge of the door, and strikes against a stop-plate fixed on the jamb at the other edge of

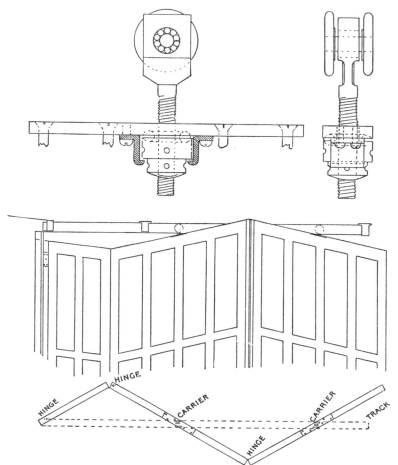

Fig. 1155.—"Coburn" Fittings for Folding Partition

the door, to prevent the door sliding too far into the recess. The pulleys in this case are of hard fibre.

The fittings for a folding partition are shown in fig. 1155. Swivel-hangers are used,

and are fixed exactly over the centres of the full-size leaves of the partition. The first leaf is one-half the width of the others, less one-half the thickness, and is hinged to the jamb as shown; all the leaves are hinged together. To find the positions of the hangers it is best to hinge the leaves and fold them together, and then strike a centre line across them. The track must be exactly over the half-leaf when this is in line (*i.e.* not folded back).

CHAPTER IV

WINDOWS AND SKY-LIGHTS

Windows consist of the glazed frames called *sashes* and *casements*, and of the frames or cases of various kinds which contain these. The sashes may be either fixed, or hinged to open like a door, or suspended by lines over pulleys and balanced by weights, or one sash may be balanced by the other sash.

The frame for the fixed sash consists of solid sides or stiles, a head-piece or lintel, and a sill, which is often made wider than the other pieces, and weathered. This frame is rebated to receive the sash, and the latter is retained in its place by a slip of wood nailed round the inside of the frame. Sometimes the outer frame is omitted, and the sash is fixed into the reveals of the brick-work or stonework, after the openings have been formed. Windows of this kind are very often used in factories, workshops, and schools, for the sake of economy. The upper part is usually fitted with a separate sash, hung on pivots a little above the centre of the height, and secured with a malleable-iron pin-stay fixed with screws to the bottom rail. A window of this kind is shown in fig. 1156. The frame consists of two side-pieces or stiles, mortised to receive the top rail or head and the bottom rail or sill; the transome is also tenoned into the stiles. The vertical section, A A, shows the details of the top rail, swivel casement, and transome. The horizontal section, B B, shows the side joint between the swivel and frame, and also the sash-bar. It is clear that the joint at *a* affords very little protection against the weather; an improvement in this respect can be obtained by means of beads planted on as at *b* and *c*, the former showing the beads above the pivot and the latter those below.

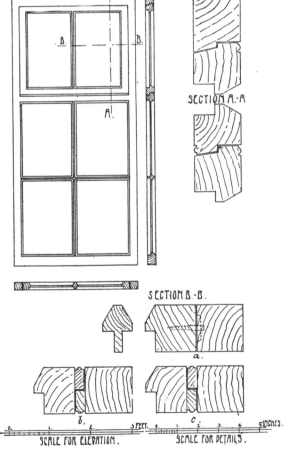

Fig. 1156.—Fixed Window with Swivel Casement

The opening casements, instead of being made to swivel, may be hung with butt hinges to the top rail of the frame, so as to open outwards; the casement and frame may then be rebated to match, and a water-tight joint can be made. Or the casement may be hung to the transome to open inwards, and in this case also the casement and

frame may be rebated. Fastenings for casements of this kind are made in almost end-less variety.

Side-hung windows are known as casement or French windows, although the latter term is now usually confined to windows extending down to the level of the floor (or

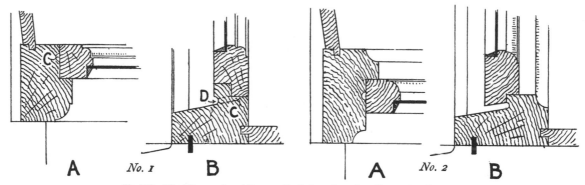

Fig. 1157.—Wood Casements and Frames: No. 1, Inward-opening; No. 2, outward-opening

nearly so), and of sufficient size to afford convenient ingress and egress. They consist of glazed sashes hung at the side to solid rebated frames, and may be arranged to open either outward or inward. The inward-opening casements are almost invariably adopted on the Continent, but in this country outward-opening casements are preferred, as they afford much better protection against driving rain; they have also the advantage of being out of the way of blinds and curtains when opened. On the other hand, the glass cannot be cleaned on both sides by a person stand-ing in the room, and when open they may be damaged by strong gusts of wind. In No. 1, fig. 1157, A is the horizontal section of the jamb and stile of an inward-opening casement, and B a vertical section of the bottom rail and sill; C C are grooves to collect the water

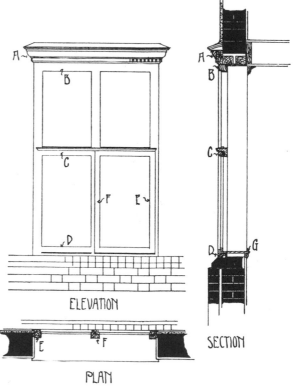

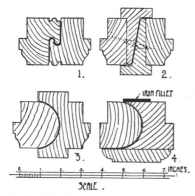

Fig. 1158.—Joints for the Meeting-stiles of Folding Casements

Fig. 1159.—Window with Outward-opening Casement and Transome Lights

driven along the joints between the casement and frame, and D is a hole (often fitted with a $\frac{1}{4}$-inch lead tube) to convey water from the grooves. A throated weather-bar is often fixed on the bottom rail, similar to that at L in fig. 1129, but smaller. The sill should be of oak, teak, or other durable hardwood, and weathered to throw off the water. An iron tongue is shown forming a weather-tight joint between the sill of the frame and the stone window-sill; sometimes the tongue is omitted, and the sill

is bedded solid on Portland-cement mortar, so that the groove in the sill is filled with the mortar (fig. 1160). In No. 2 corresponding sections of an outward-opening casement are given. In both cases the sill is grooved to receive the window-board, which may, of course, be fixed at a higher level than shown; for outward-opening casements it may be

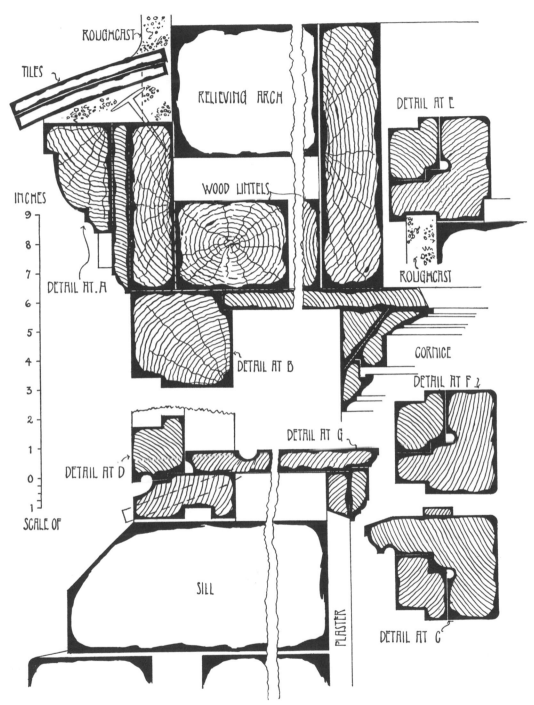

Fig. 1160.—Details of Window shown in Fig. 1159

fixed above the sill of the frame (fig. 1160), so as to abut against the bottom rail of the casement, thus forming an additional protection against the weather. The joints between the meeting stiles of casements hung folding may be simple rebates, or rebates with grooves to catch water and to dissipate the force of the wind, or formed in other ways. Four examples are given in fig. 1158; No. 1 being the hook-joint, No. 2 a joint formed with

pieces planted on the edges of the stiles, No. 3 rounded and hollowed (a joint often seen on the Continent), and No. 4 an elaboration of the latter, having a wrought-iron weather-bar outside and a wood facing inside. The best fastening for a pair of folding casements is an espagnolette bolt, but this cannot be applied to joint No. 1.

Fig. 1159 contains a plan, elevation, and section of a window with solid frame, mullion and transome, and one outward-opening wood casement in the lower part. One of the lights above the transome is fitted with an iron casement and frame let into rebates in the woodwork, the casement being hung at the bottom to open inward (known as a "fall-back"), or at the top to open outward, or pivoted at the centre (known as a "swivel"). The special feature of this window is that the opening wood casement is entirely let into the frame, and has therefore the same area of glass as the adjacent light, where the glass is let into the rebates of the framing. The details are fully shown in fig. 1160. The walls are covered outside with rough-cast, which is tucked into grooves in the jambs or stiles. A curtain-box is formed between the window-frame and the wood cornice, the soffit-board being made wide enough to form a ground for the latter. A channel for condensation water is formed in the window-board, from which a lead outlet pipe leads through the sill. Metal weather-bars and channels are sometimes fixed to the bottom rails of casements in exposed situations.

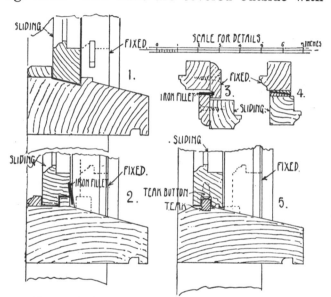

Fig. 1161.—Sections of Sills and Meeting-stiles for Sliding Casements

Sliding casements are sometimes used in cottages and other buildings where cost must be carefully considered, but if made to fit closely they are apt to jam in wet weather, and if loose they rattle in the wind. The frame is made wide enough to allow one casement to slide behind another (fig. 1161), the latter being usually fixed. The only fitting required is a small tower-bolt on the edge of the meeting-stile of the opening casement. The illustration No. 1 gives a vertical section through the sill and the bottom rail of the sliding casement, and shows the end of the fixed casement in elevation with the tenon and wedge for the bottom rail. No. 2 is a better arrangement, as the iron fillet and the wood tongue prevent rain driving into the slide-groove. No. 3 shows the meeting-stiles for the same window, the joint being made weather-tight by the iron fillet screwed to the sliding casement. No. 4 is a simpler method of forming the same joint. No. 5 shows the sill and bottom rail of a sliding sash, which will work more easily; teak buttons are let into the groove of the bottom rail and slide on a teak tongue in the sill. Brass sash-rollers may be used for the same purpose.

Suspended sashes are hung on frames provided with boxes or cases to contain the balancing weights. Fig. 1162 shows a horizontal and a vertical section of a sash window of this kind. The frame consists of sides or breasts (known as *pulley-stiles*) about 1½ inch thick, grooved down the middle for the reception of a beaded piece *o o* (No. 1), called a *parting-bead*, from its serving to part the sashes. The frame is completed by the sill below and the head above. To the outside edge of the pulley-stile the beaded pieces *e e* forming the sides of the casing or boxing are attached, and the beaded edge projects so far beyond the face of the stile as, with the parting-bead, to form the outer path or channel in which the sash slides. On the inner edge of the pulley-stile is fixed the piece *b b* (No. 1), called the inside lining or casing, to which the shutters are hinged; a back piece extending

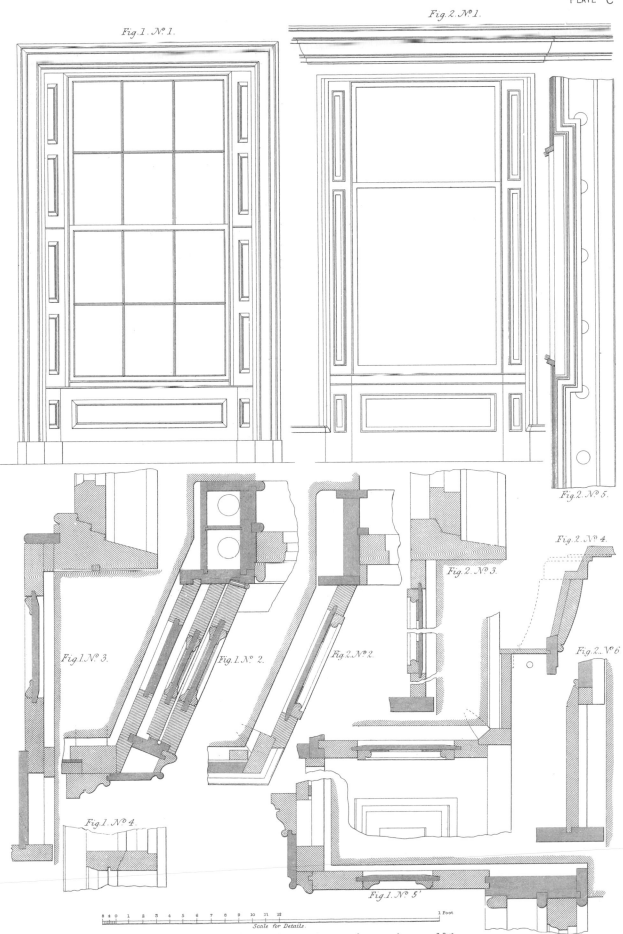

PLATE C

Fig.1.Nº.1.

Fig.2.Nº.1.

Fig.2.Nº.5.

Fig.2.Nº.4.

Fig.2.Nº.3.

Fig.1.Nº.3.

Fig.1.Nº.2.

Fig.2.Nº.2.

Fig.2.Nº.6

Fig.1.Nº.4.

Fig.1.Nº.5'

Scale for Details.

1 Foot

Scale for Elevation.

7 Feet

SASH WINDOWS AND THEIR FINISHINGS

between the inside lining and outside piece *e*, parallel to the pulley-stile, is added to complete the case or boxing, and the box has sometimes, and should always have, a division or "parting slip" in the centre to separate the weights of the upper and lower sashes, as shown in No. 1. The path for the inner sash is formed by a slip or stop bead, fixed to the stiles by nails or, preferably, by screws. In the lower end of the path of the outer sash a hole is cut in the pulley-stile sufficiently large to admit the weights, so that the sashes may be hung after the frames are fixed, and the lines repaired at any time. This is called the pocket, and it is covered by a piece of wood attached by screws.

The sashes themselves consist of an outer frame, which is composed of stiles and rails. The bottom rail of the lower sash is deeper than the others, and is throated to prevent the water from driving under it. It is a good plan to make this rail about 5 inches deep, and to make the beaded lining of nearly the same height, so that the lower sash can be raised sufficiently to admit air at the meeting-stiles, without rising above the bottom lining; this affords a simple method of admitting air without creating an objectionable draught. The meeting rails of the upper and lower sashes are, in good work, made wider than the others, and fitted together in the manner shown at A, No. 2. The horizontal and vertical bars, which divide the sashes into panes, are termed *sash-bars*. The vertical bars, like the stiles in framing, extend, in single pieces, between the rails of the sash, the horizontal bars being scribed to fit, and dowelled together through the vertical bars. The rails and horizontal sash-bars are tenoned through the stiles and secured with wedges, and wood pins are usually inserted at the angles of each sash to give additional security.

The fittings of a window consist of the boxings for the shutters (if there be any), the linings, the shutters with their back-flaps, and the architraves or other finishings of the opening in the apartment. All these parts are exhibited in No. 1 in horizontal section. The boxings are formed in the space between the inside lining of the sash frame and the framed ground. The back of the recess is sometimes plastered; but, in better work, it is covered with a framed lining *a*, called a back lining. This has generally bead and flush panels, and is fitted between the inside lining *b* and the framed ground *c*, and is generally tongued into both. The shutters, *d*, are framed as doors, and panelled and moulded in the same manner. They are hinged to the inside lining. The back-flaps are generally lighter than the shutters, and are sometimes framed and moulded, so that the whole exposed surface shall present the same appearance when the shutters are closed; but they more frequently consist of bead and flush framing.

The woodwork M M (No. 2), which extends from the window sill to the skirting, is

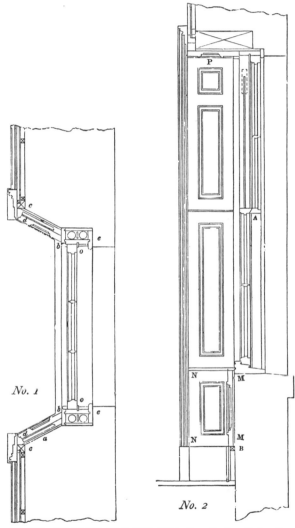

No. 1

No. 2

Fig. 1162.—Sash-window: No. 1, Horizontal section; No. 2, vertical section

called the *breast* lining; and that on the side of the recess, N N, extending from the bottom of the shutters to the skirting, the *elbow* lining. The ceiling of the window recess P is also formed of wood, and is termed the *soffit* lining. These linings are all framed and moulded to correspond with the doors and other framed work of the room. The margin of the window opening is finished with architraves or other ornamental appliances, in the same manner as the doors. The wall in this example is battened to receive the laths and plaster.

PLATE C.—Fig. 1, No. 1, is the elevation of a sashed window with its finishings. Fig. 1, No. 2, is an enlarged section of one of the window jambs, showing in detail part of the lower sash, the pulley-piece boxing, shutter boxing, back lining, architrave, and shutters. Fig. 1, No. 3, is a vertical section through the lower part of the window and window breast, showing the rail of lower sash, the sill of the window frame, the breast lining and skirting. The bead is in this case stuck on the sill, but usually it is a separate member planted on. The upper part of the window is shown at the bottom of the plate in fig. 1, No. 5. It shows the rail of the upper sash, the lintel of the window frame, the soffit lining, and the architrave. Fig. 1, No. 4, is a section through the meeting rails of the upper and lower sashes, showing the nature of the rebate, or check. Fig. 2, No. 1, shows a sashed window with flush breast and elbows. Fig. 2, Nos. 2, 3, and 4, show details of the mock shutter (or panelled lining), breast, and soffit to a larger scale. The breast is finished flush, as no projecting sill is required where mock shutters are used. Fig. 2, No. 5, is a view looking up to the cornice over the window. A curtain box is formed in the cornice, and the

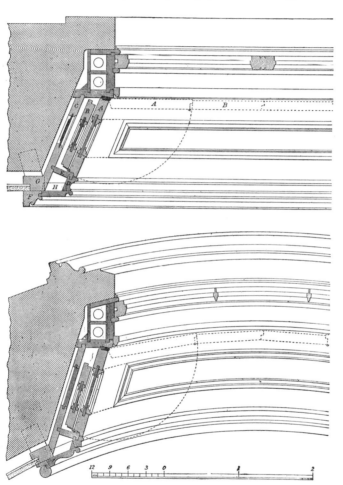

Fig. 1163.—Sash-window with Internal Shutters

bed mouldings only returned round it. If venetian blinds are used the lintel should be kept 10 inches above the daylight to allow the blinds to be drawn up clear of the sash. Fig. 2, No. 6, shows the skirting, which is finished flush against the lining of the window.

No. 1, fig. 1163, shows the finishing of a window, looking upwards towards the soffit. The dotted lines A B show the shutters and back-flap when closed; and the shaded parts, A B, the shutter and back-flap when folded into its boxing; C H D E is the boxing and back lining; G the ground; and F the architrave. No. 2 shows the soffit, shutters, and finishings of a window in a circular wall.

Plate CI contains a plan, elevation, section, and details of the internal finishings of a sash-window in the King's Gallery, Kensington Palace, and is an excellent example of eighteenth-century work. The sash-bars are stronger than those now generally used, and the meeting rails are below the centre of the window, this arrangement being probably

adopted either to obtain an odd number of squares or to render the meeting rails more accessible. The finishings are well designed, and include a window-seat with panelled front, panelled window-linings, and moulded and carved architraves resting on pilasters; the moulded capping of the panelled dado is continued to form the nosing of the window-board at the back of the seat.

Sash-window on Circle.—Plate CII, fig. 1: No. 1 is the plan, and No. 2 part of the elevation of a sash-window with diamond panes, the plan being a segment of a circle. In No. 1, A are the jambs, B the sash, C the pulley-stile, E F the window-sill, D the inside lining. The centre lines of the divisions for the panes are set out on the plan of the lower rail of the sash, at cd, ef; the thickness of the bars at kl, o, &c.; and the sites of the crossings or intersections of the bars at hi, n, &c. The heights are set out in a similar manner on No. 2, and from these data the elevation can be drawn, as in the manner already familiar to the reader. Fig. 2 is an elevation of the lower sash on the stretch-out; and fig. 3 the stretch-out of the diagonal bar A D, showing the twist occasioned by the curvature.

To find the Mould for the Curvature of the Diagonal Sash-bar.—In fig. 4, let A B E D be the plan of the bottom rail, and let it be divided into any number of equal parts— 1, 2, 3, 4, &c. Through these draw ordinates at right angles to the chord line A B, meeting it in $abcd$, and produced beyond it to meet the chord line A C of the diagonal bar in $lmnop$. Through these points of intersection draw ordinates perpendicular to A C, and make them equal to the corresponding ordinates of A B, as lqv to $af1$, mrw to $bg2$, &c.

Fig. 5 shows the section of a sash-bar, full size, with a dowel hole A; and underneath is shown the mitring of the bars B D C E, and the dowel A A, all full size.

Circular-headed Sash on Circle—PLATE CIII.—The figures in this plate illustrate the framing of a circular-headed sash in a circular wall. Fig. 1, No. 1, is a plan of the window, and No. 2 an elevation of the circular-headed sash. In No. 1, A A are the jambs, S the outside lining, B C the upper sash, G G the pulley-stiles, H the inside lining, E the parting bead, F the stop bead, or batten rod, as it is sometimes called, and M the sill.

To find the Veneer for the Arch-bar K L M, *called the Cot-bar or Chord-bar.*—Set out the stretch-out of the arc K L M, fig. 1, on the line A B, fig. 4, and draw lines from the divisions in the arc to any chord line, as N o, No. 1; then make the ordinate $c a$ D, fig. 4, equal to ozP, No. 1, $3be$ equal to rty, and so on; then G E D F H, fig. 4, will be the veneer for the arch-bar.

To find the Mould for the Radial Bars.—From P in No. 1 draw P R a tangent to the curve; and on it draw lines from the divisions in the radial bars F H, E G, and produce them to cross the plan of the sash; then transfer the ordinates R hi, jkl, &c., to H rs, $3tu$, &c., in No. 2, and mn, op, &c., to G w, $3x$, &c., and the moulds L G, F s, of the bars E G, F H, will be obtained.

To find the Face Mould for the Circular Outside Lining.—The dotted line $aklmnop$, fig. 1, No. 2, shows the lower edge of the lining; and lines drawn through these points perpendicular to A C, cut the line sg (No. 1), in $abcdefg$. Transfer these on the stretch-out to the line A B, fig. 5, and draw ordinates perpendicular to A B, on which set up the corresponding heights from No. 2, as bk to bg, cl to ch, dm to dk, &c.

To obtain the moulds for the head of the sash-frame apply the stretch-out of the outside of the arch in No. 2 to the base line A B in fig. 2, and set out on the ordinates drawn through the divisions, the corresponding ordinates from the chord I K in No. 1.

To obtain the mould for the underside of the sash, fig. 3, set out the divisions of the underside of the arch in fig. 1, No. 2, along the base line A B in fig. 3, and proceed in the same manner as above, but setting out the ordinates from the chord line L M in fig. 1, No. 1,

Fig. 1, No. 3, shows the first division of the sash-frame A N in No. 2, and the plan No. 1; the thickness of stuff required to work it out of the solid is shown at E N. The joint at N, No. 2, and *k h*, No. 3, is shown at *b c* in fig. 2, and C F in fig. 3.

Figs. 6, 7, and 8 are sections through the sash-frame and sash at three points: 1st, above the springing; 2nd, at *f*; and 3rd, at the centre. The part corresponding to the pulley-stile is now divided into two pieces, B and C, and the parting bead *b* is inserted between them. *a* is the outer, and *c* the inner lining; the latter beaded and grooved for the reception of the soffit lining. A is the sash, the rebate and mould not being shown.

Combined Sash-and-casement Window.—A combination of a sash and casement is shown in fig. 1164. The window consists of two sashes hung in the usual way, but the lower sash is much higher than the upper. The stiles of the lower sash have tongues sliding in grooves

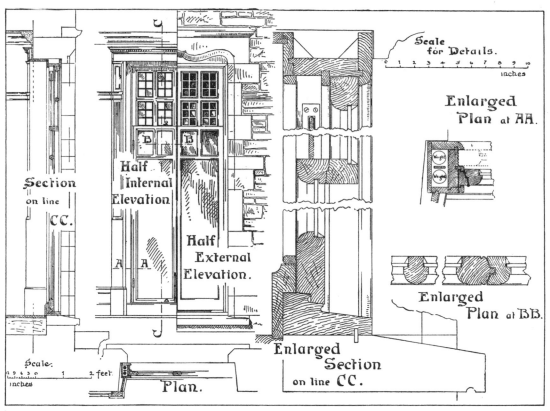

Fig. 1164.—Combined Sash-and-casement Window

in the parting bead and inside bead. To these stiles the French windows are hung. It is obvious that, when the French windows are closed, they and the stiles to which they are hung can be raised together like an ordinary sash. When the lower sash is shut, the French windows can be opened, the sliding stiles being kept in position by the shoulders of the beads. This is not a patented arrangement; the writer first saw it in the house of a friend in North Wales, and was told that it had been copied from an old window in the neighbourhood. It is well adapted for the upper stories of a building, as it allows the glass to be cleaned on both sides without danger. There are many patented devices for attaining the same end, but these cannot be considered here.

Sash-cords or *lines* may be of jute, flax, or twine, the last being the most costly; each kind is made in seven strengths or weights. They are sold in "knots" of 12 yards, or per gross yards. Sometimes copper or steel chains or ropes are used instead of fibre.

Sash-weights are usually of cast-iron, but where space is limited, or where the sashes are very heavy, lead weights are used. The weights must balance the glazed sash to which they are attached.

Sometimes one sash is hung to balance the other, the weights being omitted; as both sashes must be opened together, the bottom rail of the lower sash should be made deeper than usual, as already described, so that air can be admitted at the meeting rails and top without raising the bottom rail above the inside bead.

Sash-pulleys or *axle-pulleys*, over which the sash-cords run, are made in sizes from $1\frac{1}{2}$ to $3\frac{1}{2}$ inches in diameter, those above 2 inches being only used for heavy sashes; the face-plates and wheels may be of iron or brass, and the bushes of iron, brass, or gun-metal.

Sash-fasteners are usually fixed on the top of the meeting rails; they are of two main types, the screw and the quadrant, and ought to be so designed that the sashes are drawn

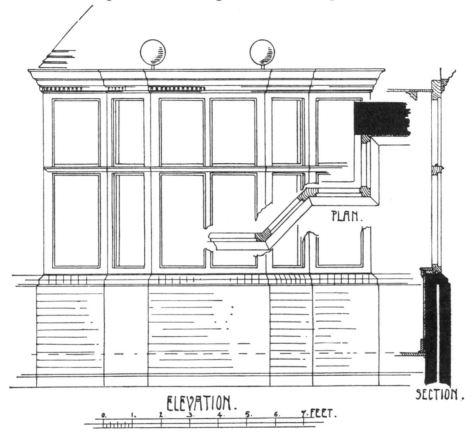

PLAN.

ELEVATION.

SECTION.

0.   1.   2.   3.   4.   5.   6.   7. FEET.

Fig. 1165.—Bay-window with Iron Casements

together by closing the fasteners (in order to reduce rattling) and that the fasteners cannot be opened from the outside without breaking the glass. Other fasteners, for securing the top and bottom sashes independently, are made with spring bolts which operate automatically when the sashes are shut.

Sash-acorns are small brass studs which screw into threaded sockets let into the inner face of the upper sash about 6 inches above the meeting rail. Either sash can be opened to this distance, and no further, unless the acorn is unscrewed. They afford some protection against unskilful burglars.

Sash-lifts are usually of stamped brass, bent in the form of a hook, and screwed to the bottom rail to facilitate the raising of the lower sash.

Bay-windows.—A bay with brick plinth and wood framing, forming part of a house now being built by the writer at Byfleet, Surrey, is shown in fig. 1165. The framing is $3\frac{3}{4}$ inches thick, and is rebated to receive wrought-iron casements and leaded lights. The angle mullions are worked out of the solid, and the details generally follow those given in figs. 1159 and 1160. The sills are of oak, and the other framing of red deal,

The balls are of wood covered with lead, and supported on copper dowels. A lead flat is formed over the projecting portion in the middle of the bay, the remainder of the roof being tiled.

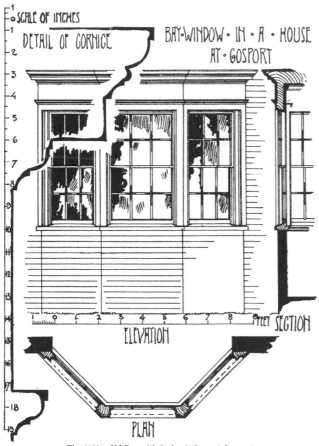

Fig. 1166.—Old Bay with Sash-windows at Gosport

Fig. 1166 shows a small bay at Gosport, fitted with sash-windows. The details of construction are not shown, but will be understood by the intelligent reader.

Sky-lights.—The ordinary sky-light for a sloping roof is shown in fig. 1167. The rafters are trimmed to receive the frame, which rises above the slates or tiles, and is rebated at the top to form a tongue which fits into a groove in the underside of the sky-light itself. The sky-light overhangs the frame, and is throated around; the bottom rail is thinner than the stiles and top rail, to allow the glass to go over it and form a drip. Sometimes the glass is kept clear of the bottom rail, so that condensation water can escape between the glass and the rail; but as this opening allows driving rain to enter, it is sometimes omitted and a condensation groove formed in the upper part of the rail. The sky-light is hinged to the frame at the top, and is usually fitted with a pin-stay, so that it can be fixed in any desired position.

For the general arrangement of top-lights in the roofs of weaving sheds, see fig. 722, page 18, Vol. II. These lights are made from 2 inches in thickness upwards, according to the size. They are often made in sections, the width being equal to the distance between the centres of the cast-iron standards to which they are bolted or screwed. The joints over the standards may be cross-tongued, and, in addition to this, $\frac{3}{4}$-inch chamfered facings are usually planted on to cover the joints; a similar facing covers the joint between the top-light and the ridge-piece.

In the sky-light (fig. 1168), of which No. 1 is the plan and No. 2 the elevation, it is required to find the length and backing of the hip. Let A B be the seat of the hip; erect

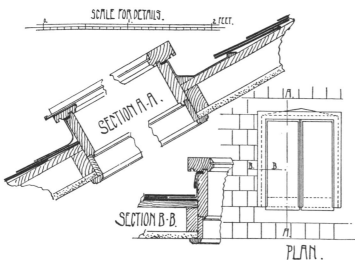

Fig. 1167.—Sky-light in Sloping Roof

the perpendicular A C, and make it equal to the internal height of the sky-light, and draw B C, which is the line of the underside of the hip; the dotted line *g h* shows its upper side. To find the backing: from any point in B C, as *n*, draw perpendicular to B C a line *n* F meeting A B

in F, and through F draw a line at right angles to A B, meeting the sides of the sky-light in D and E.   Then from F as a centre, and with F n as radius, cut the line A B in m, and join D m, E m.   The angle D m E is the backing of the hip, and the bevel k m l will give the angle of backing when applied to the vertical side of the hip bar.

In fig. 1169, in which No. 1 is the plan, and No. 2 the elevation of a sky-light with curved bars, to find the hip, let A B be the seat of the centre bar, and D E the seat of the hip. Through any divisions 1 2 3 4 c of the rib over A B, draw lines at right angles to A B, and produce them to meet E D in p o n m D.   From these points draw lines perpendicular to E D, and set up on them the corre-sponding heights from A B, as l 1 in p 1, k 2 in o 2, &c.

In the irregular octagonal sky-light (fig. 1170) the length and backing of the hips are found, as in fig. 1168, by drawing A C per-pendicular to A B, and setting up on it the height of the sky-light in A C, then drawing B C.   The back-ing D h E is found by drawing g F perpendicular to B C from any part of B C, and through F drawing D E at right angles to A B to meet the adjacent sides of the sky-light in D E, then making F h equal to F g, and joining D h, E h.

In the octagonal sky-light with curved ribs (fig. 1171) the process of finding the hips is exactly similar to that employed in fig. 1169.

PLATE CIV.—Nos. 1, 2, and 3, fig. 1, are the plan, side ele-vation, and end elevation of an irregular octagonal sky-light, and Nos. 1, 2, and 3, fig. 2, the plan, side elevation, and end elevation of an elliptical sky-light, neither of which requires detailed descrip-tion.

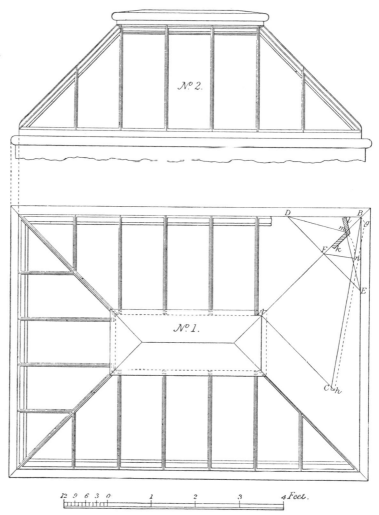

Fig. 1168.—Rectangular Sky-light with Straight Ribs

In fig. 3, No. 1 is the plan, No. 2 the end elevation, and No. 3 the side elevation of an elliptical domical sky-light.   The section of the sky-light on the minor axis is a circular segment, as seen in No. 2.   To find the ribs (fig. 4), describe the quadrant A B, and in C B make the height D B equal to the height of the segment in No. 2; draw E D, and make E L equal to the length of the rib over the minor axis, and draw C L to find the bevel L w of the end.   Divide the arc E L into any number of parts, and through them draw lines per-pendicular to A C, and produce them indefinitely; draw also, from the lower end of the rib, the line m K perpendicular to A C.   Then from D as a centre, with the length of the longest semi-diameter of the ellipse as radius, cut the line m K in the point K, and draw D K, and produce it to n to meet the perpendicular A n from A.   Then the line D n will be the semi-axis major of an ellipse, as A C in fig. 5, and the segment of it formed by the lines L u, m K, in fig. 4 will be the rib standing over the semi-axis major.   But all the ribs may be drawn by

ordinates thus: From D, fig. 4, as a centre, and with the lengths of the several ribs as radii, cut the line m K in H, G, F, and through these draw lines from D, meeting A n. Then the points where these lines are crossed by the perpendiculars to A C, passing through the divisions 1 2 3 4 L in the arc E L, are the places of corresponding ordinates by which the curves may be drawn, as D u t K at D u t o in fig. 5, D u t H at D u t o in fig. 6, D u t G at D u t o in fig. 7, and D u t F at D u t o in fig. 8.

PLATE CV.—In fig. 1, No. 1 is the plan and No. 2 the elevation of an octagonal sky-light. No. 3 is one of the sides laid over on the horizontal plane of projection. In fig. 2, No. 1 is the projection of a portion of the inside of the sky-light looking up, and No. 2 is an elevation of a portion of the interior corresponding to the last. The mode of finding the lengths and backings of the hips and ribs is developed in the lower half of the plan. First, the hip B D. Draw D E perpendicular to B D and equal to the vertical height, and join B E. From any point b in B E let fall a perpendicular meeting B D in c; make c a equal to c b, and join c a and A a, and produce the latter to T. Then c a T is the bevel for the backing of the hip to be applied to the vertical side of the rib.

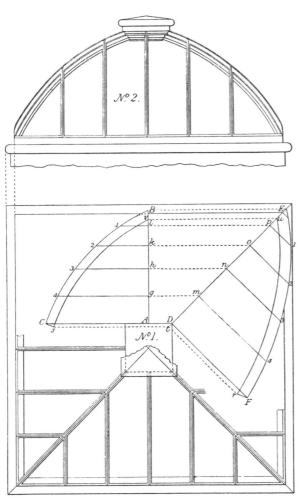

Fig. 1169.—Square Sky-light with Curved Ribs

It will be seen that the rib K I is found in a similar manner from the right-angled triangle K I L, of which the hypotenuse K L is the length of the rib as before.

In obtaining the ribs on the lower side of the octagon a compendious method is adopted. Let P Q be the seat of the hip, and N O, G F the seats of any other ribs; on F G construct the right-angled triangle F G H as before, and from any point R in the hypotenuse draw R S parallel to H G, and R e at right angles to F H. From R as a centre with any radius describe a circle as d e f g, and through e and f draw lines parallel to H G. At the points where these cut the seats of the ribs erect perpendiculars, as at n m, k l, p o, i h, and intersect them by tangents from the circle parallel to H G, as d l m, g h o; then join m s and o s, and we obtain R S o and R S m as the bevels for the backing of the rib P Q, and in like manner the backing of any other rib is obtained.

Fig. 3, No. 1, shows the rib at F G in fig. 1, No. 1, to a larger scale, and No. 2 shows a hip rib, the angle of backing Q D S being the same as A a C in fig. 1, No. 1. No. 3 shows the common bar corresponding to the line I K in fig. 1, No. 1, the angle P F o being the same as w r x in the latter figure. No. 4 is the hip as seen in fig. 2, Nos. 1 and 2, which is cut out of thicker stuff, and No. 5 one of the horizontal bars. The manner of finding the mouldings of the angle bars and ribs, as exemplified in this figure, has been already described in detail; the same letters refer to the same parts in all the mouldings, by which their correspondence can be readily traced.

Figs. 1 and 2, Plate CVI, are sections of a sky-light in a flat roof. There are louvre-

PLATE CI

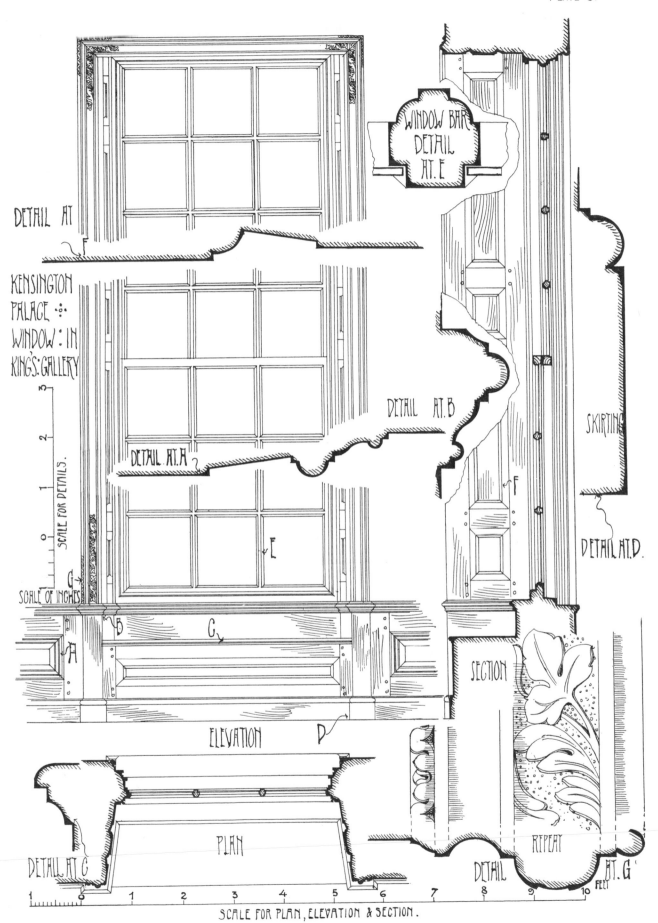

DETAIL AT

KENSINGTON
PALACE ❖
WINDOW ❖ IN
KING'S ❖ GALLERY

WINDOW BAR
DETAIL
AT. E

DETAIL AT B

DETAIL AT A

SCALE FOR DETAILS.

SKIRTING

DETAIL AT D.

SCALE OF INCHES

G

A

B

C

E

D

ELEVATION

SECTION

REPEAT

DETAIL AT C

PLAN

DETAIL AT G

FEET

1   0   1   2   3   4   5   6   7   8   9   10

SCALE FOR PLAN, ELEVATION & SECTION.

WINDOW IN KING'S GALLERY, KENSINGTON PALACE

PLATE CII

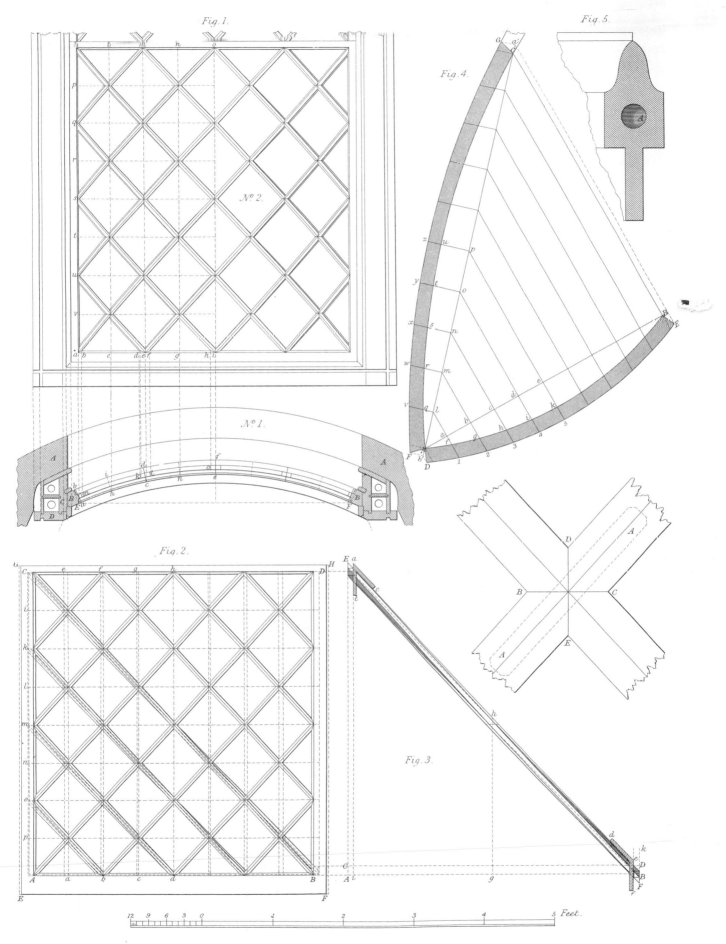

Fig.1.

Fig.5.

Fig.4.

No.2.

No.1.

Fig.2.

Fig.3.

12 9 6 3 0   1   2   3   4   5 Feet.

SASH WINDOW ON CIRCLE

PLATE CIII

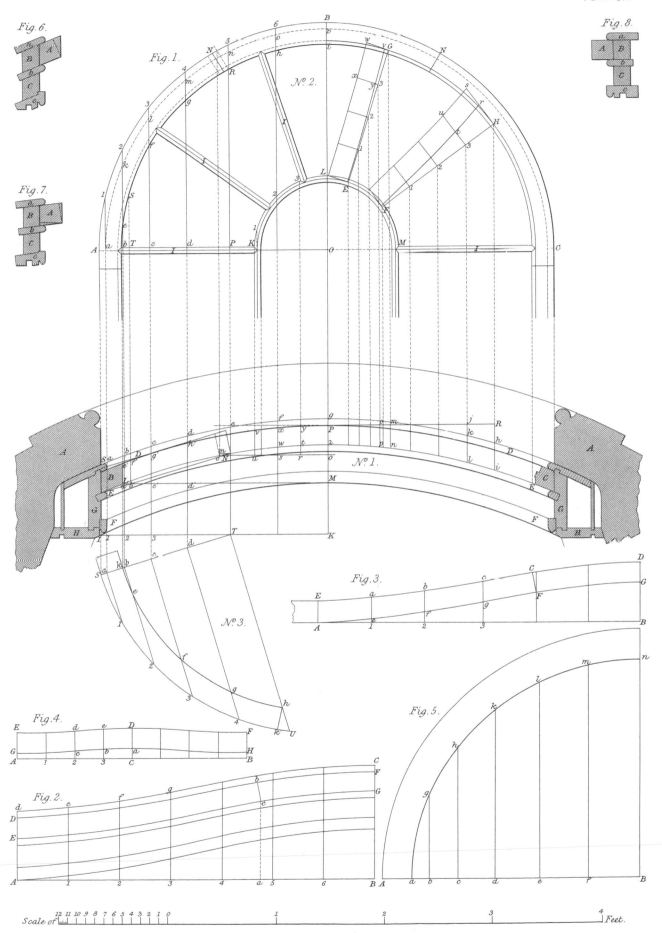

Fig.6.

Fig.7.

Fig.1.

Fig.8.

Nº 2.

Nº 1.

Nº 3.

Fig.3.

Fig.4.

Fig.5.

Fig.2.

Scale of 12 11 10 9 8 7 6 5 4 3 2 1 0 ⟋1 ⟋2 ⟋3 ⟋4 Feet.

CIRCULAR-HEADED SASH ON CIRCLE

PLATE CIV

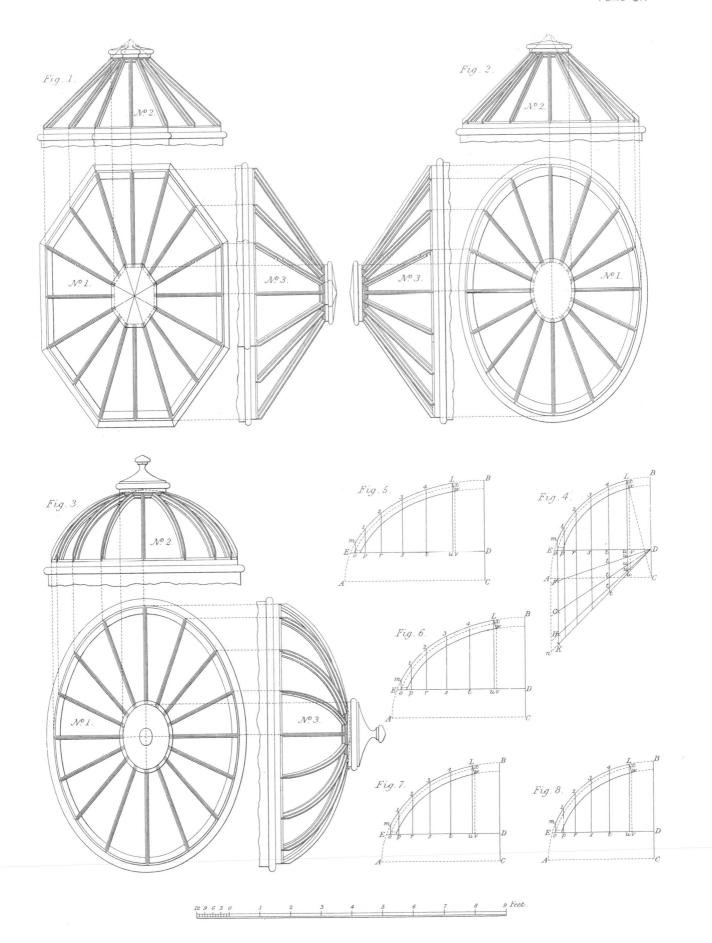

Fig. 1.

Fig. 2.

Fig. 3.

Fig. 4.

Fig. 5.

Fig. 6.

Fig. 7.

Fig. 8.

SKYLIGHTS

PLATE CV

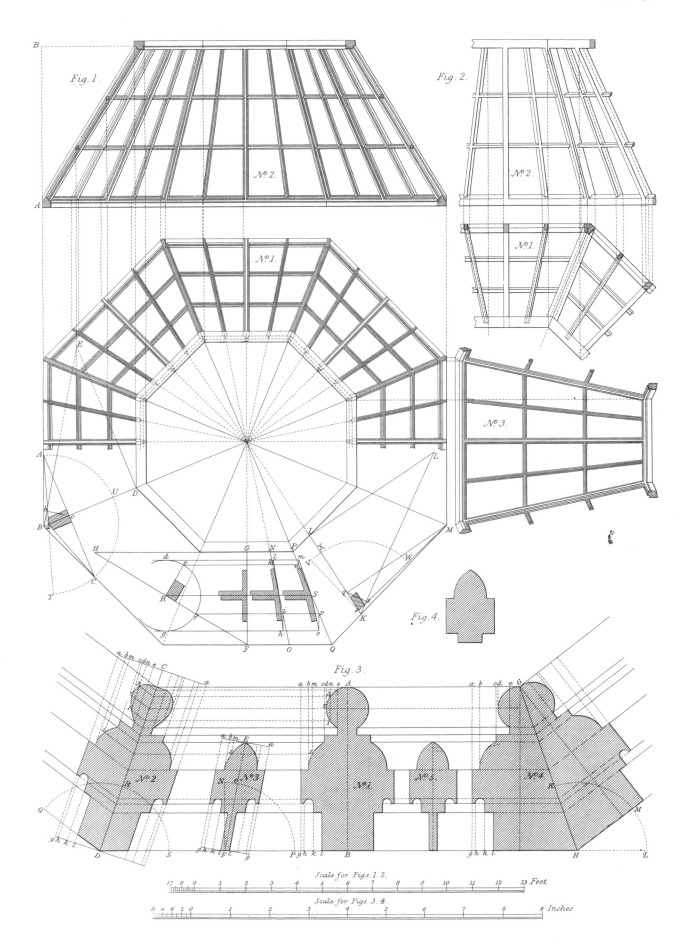

Fig. 1.

Nº 2.

Nº 1.

Fig. 2.

Nº 2.

Nº 1.

Nº 3.

Fig. 4.

Fig. 3.

Nº 2.   Nº 3.   Nº 1.   Nº 5.   Nº 4.

Scale for Figs. 1. 2.

12 6 0   1   2   3   4   5   6   7   8   9   10   11   12   13 Feet

Scale for Figs. 3. 4.

8 6 4 2 0   1   2   3   4   5   6   7   8   9 Inches.

OCTAGONAL SKYLIGHTS

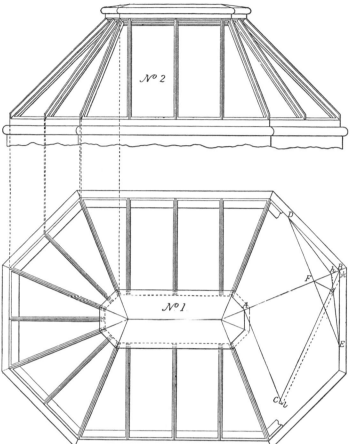

Fig. 1170.—Irregular Octagonal Sky-light with Straight Ribs

board ventilators a, a in each end, which may be kept open or covered inside by an ornamental perforated board, as shown at $b\,b$. The glass g rests at the sides on wood astragals $c\,c$ (one of which is shown to a larger scale in fig. 3), and is firmly secured with putty, while the top of the pane is fitted into a zinc ridge $d\,d$ (see fig. 4). Small gutters are made in this ridge, $e\,e$, fig. 4, for retaining any water that may be blown in over the glass; the water is led to the gutters $f\,f$ in the astragal, fig. 3, and is discharged at the sill. The projection of the glass g, as shown in fig. 5, and the hollow in the top of the sill, prevent the rain-water being blown back, while the glass g being raised from the sill allows condensed water to escape under it.

Fig. 5 shows how the sill can be fixed to the framing without damaging the lead apron g^2. When the framing is set up round the opening for the window, a wood fillet j should be placed round the frame h, as indicated by dotted line $i\,i$; the upstand of lead apron should be dressed against this fillet, after which the fillet is removed, and the sill with groove k in bottom is set on and nailed inside of apron. Fig. 6 is a plan of the same window "looking down", and shows the cope mm at each end, and the ridge $d\,d$.

The sky-lights which have been shown in figs. 1168 to 1171, and in Plates CIV to CVI, are raised above the roofs, and are often known as dome-lights. The same methods of construction can be applied to the roofs of greenhouses, &c., but the flats should be omitted and the bars continued up to the ridge-piece.

Dormers.—Fig. 7, Plate CVI, is an elevation, and fig. 8 a section of a dormer window, the frame of which is made of cast-

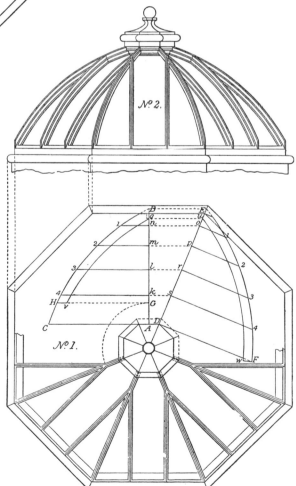

Fig. 1171.—Octagonal Sky-light with Curved Ribs

iron, and the sash nnn of wood. It is suitable for a mansard or other steep pitched roof. The spars are bridled, as shown at oo; the frame of window pp is fixed to the sill o and to the sarking q by means of screws. The ribs of roof rr are put under the frame for the purpose of fixing the laths. The sash nn is made to open, and is hung on pivots at the centre; it may also be hinged at the bottom or top.

Fig. 1172 contains an elevation and section of a wood dormer with details to a larger scale. There are two opening casements of wood, which are let entirely into the jambs,

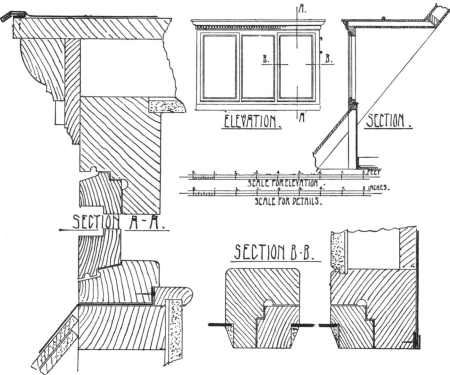

Fig. 1172.—Dormer with two opening Casements

mullions, and top rail of the framing. The central light is glazed directly to the framing, a false bottom rail being inserted to match the rails of the opening casements. The sills are of oak, and the section A A shows the method of fixing the lead apron. Other designs for dormers are given in figs. 789 and 790, page 73, Vol. II, and in Plates XXI and XXII.

Borrowed lights are glazed sashes in internal walls, and may be either fixed or made to open as sashes or casements. As the construction does not differ materially from that of windows, illustrations are unnecessary.

CHAPTER V

SKIRTING, WALL-PANELLING, CORNICES, PEWS, &c.

Skirting is the name commonly given to the square or moulded board fixed along the bottom of an internal plastered wall or partition. It is sometimes known as a base, plinth, or wash-board. Some examples have already been given in the chapters on doors and windows. Fig. 1173 contains other examples, and shows various methods of fixing. Skirtings not exceeding 9 inches in depth are generally made in one piece, as in No. 1, unless the moulding has a considerable projection, in which case the moulding is run on a separate piece, known as the "base-moulding" and jointed to the lower portion, or

"skirting board", as shown in No. 2. Nos. 3 and 4 are other examples of skirting in two pieces, and No. 5 shows a deeper skirting in three pieces.

The joint between the skirting and floor-boards is sometimes a simple "butt" joint, as shown in No. 1, but as skirtings are now usually fixed before the plaster is thoroughly dry, shrinkage occurs in them as well as in the floor-joists, and an unsightly opening is left along the bottom of the skirting; to conceal this, a rounded or hollowed fillet is nailed to the flooring, as shown at *f*. The same object can be attained by fixing a ground (of a smaller thickness than the floor-boards) to the joists, so that a small pocket is formed to receive the lower edge of the skirting (Nos. 2, 3, and 4). Where two thicknesses of floor-boards are laid, the lower is continued to the wall, and the upper stopped at the face of the skirting. Occasionally a tongue-and-groove joint is formed (No. 5), but this is a laborious method and is now less frequently adopted than the others. When the simple butt joint shown in No. 1 is adopted, the operation of scribing should be performed, to accommodate the outline of its lower edge to any inequalities which may exist on the surface of the floor. In doing this, the skirting board, having its upper edge worked with perfect precision, is applied to its place, with its lower edge

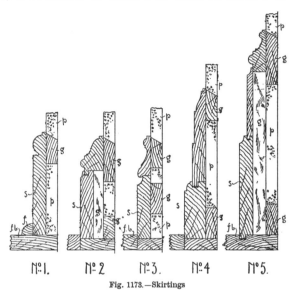

Fig. 1173.—Skirtings

supported at a convenient distance above the floor, and so arranged by propping it up at one end or the other that its upper edge is perfectly level. Then to mark on its lower edge a line that will perfectly coincide with any irregularity which may exist in the floor, a pair of strong compasses is taken and opened to the greatest distance that the lowest edge of the skirting is from the floor throughout its length. The outer point of the compasses is then drawn along the floor, and the other point pressed against the skirting board, so as to mark a line which will be exactly parallel to the surface of the floor; and the board is then cut to this line. It is, of course, essential that the upper edges of all the skirting boards of the room or apartment should be adjusted to the same level line when this scribing is done.

The illustrations show the methods of fixing the skirtings to rough grounds, *g*. The chief point to be observed is that each member should be fixed along one edge only,

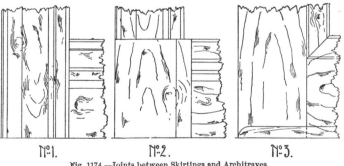

Fig. 1174.—Joints between Skirtings and Architraves

the other edge being left free to allow for expansion and contraction. Where the backs of the members, of which the skirting is built, are not in the same plane, vertical grounds may be fixed to the horizontal grounds, as in No. 5, after the plastering has been floated down to the floor. In all these illustrations, *s* is the skirting, *g* the ground, *p* the plaster, and *fb* the floor-boards.

The joints at the external angles of skirting are formed by mitring, and those at internal angles by mitring or scribing. Those between skirtings and architraves may be formed, as in No. 1, fig. 1174, by running the architrave down to the floor and "butting" the skirting against it. A more finished arrangement is shown in No. 2, the skirting

"butting" against a plinth or block fixed at the foot of the architrave. Sometimes the same mould is run on the skirting and architrave, and the two are mitred together as shown in No. 3.

Match-boarding is sometimes used as a wall-covering. It ought to be in narrow widths, so that the shrinkage will be distributed over a large number of joints. The boards are usually placed vertically and nailed to horizontal grounds from 2 to 4 feet apart. They may be tongued and grooved or rebated, and either beaded or V-jointed, but occasionally the whole surface is moulded, as in No. 1, fig. 1175, or plain and moulded boards are fixed alternately, as in No. 2. Secret-nailing should be adopted where possible, particularly if the boards are to be varnished or polished. Dadoes may also be formed with vertical match-boarding, finished at the top with a moulded capping level with the window-board; they are usually from 2 feet to 3 feet high in houses, and 4 feet high in schools. The capping may take the form of a rail

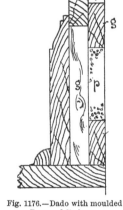

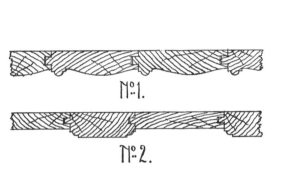

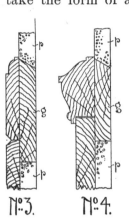

Fig. 1175.—Match-boarded Dadoes, &c.

Fig. 1176.—Dado with moulded Base and Surbase

of the same thickness as the boarding, as in No. 3, fig. 1175, or may project as in No. 4. More elaborate dadoes are sometimes formed with horizontal cross-tongued boards, and

Fig. 1177.—Panelled Hall and Mantel-piece in a Surrey House

moulded plinths (or "bases") and capping (or "surbases"), as shown in fig. 1176. Instead of the cross-tongued boarding, panelled and moulded framing may be used, the thickness of the framing being, as a rule, not less than $\frac{7}{8}$ inch.

Plate CVII contains an elevation and details of the wall-panelling in Queen Caroline's chamber, Kensington Palace. The great width of the panels, which are raised as shown in the detail at B, is a noteworthy feature of the design. The spaces over the windows are occupied by panelled framing, which appears to be planted on the face of the main framing. The architrave over the central window, and the cornice at the level of the springing, are elaborately carved. The sash-bars, D, are much stouter than in many modern windows.

A simpler type of panelling in a house in Surrey, built from the writer's designs, is shown in fig. 1177. The framing is $\frac{7}{8}$ inch in thickness, and is carried up to the height of the doors; a rounded fillet is planted to the floor to conceal the joint between the panelling and flooring. The panelling is fixed to $\frac{3}{4}$-inch rough grounds, and the walls are plastered behind the panelling. It is a good plan to paint the back of all panelling and skirting in order to reduce or prevent the absorption of moisture from damp plaster. Another example of wall-panelling, carried out in mahogany, is given in fig. 1070.

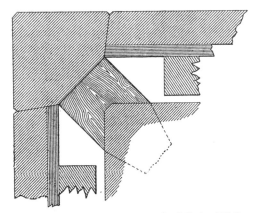

Fig. 1178.—Staff-bead fixed to Battened and Plastered Wall

Boarded partitions or "ceilings" are often used in cottages and in attics, workshops, and other places. The simplest are formed with tongued and grooved match-boarding, fixed vertically and nailed to transverse pieces known as backstays, which are framed into uprights, about $2\frac{3}{4}$ inches by $2\frac{1}{4}$ inches or larger. These uprights are fixed at all angles and to form the jambs of doorways. Sometimes the boarding is double, the backstays being concealed between the two thicknesses. Panelled framing is often used instead of match-boarding, the upper panels being in some cases glazed.

Angle-beads or *staff-beads* are pieces of wood fixed vertically to the salient angles of internal plastered walls and partitions, where cemented angles are not run. They are fixed to plugs driven into the joints of the brickwork, and may have square angles as shown in fig. 1178, or may be in section about three-quarters of a circle.

Chair-rails or *dado-rails* are moulded rails fixed to horizontal grounds at or near the height of the top of a chair—about 3 feet from the floor,—but the height may be varied to suit the level of the window-board. A rail of this kind, carried around the room at the level of the window-board, is shown in Plate

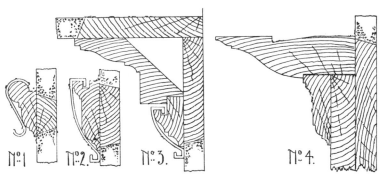

Fig. 1179.—Picture-moulds, Cornice, and Plate-rail

LXXXVIII. The thickness is the same as that of the mantel-piece framing and door-architraves, against which it "butts", namely $\frac{7}{8}$ inch; the depth is $2\frac{3}{4}$ inches.

Picture-moulds are horizontal mouldings fixed around rooms, usually a short distance below the ceilings. As the width of a large wall-paper frieze is 21 inches, it is a good plan to leave a clear space of 21 inches between the picture-mould and cornice; in low rooms, the space may be 18, 15, $10\frac{1}{2}$, or 7 inches, to suit the narrower widths of frieze. The stock-pattern picture-hook is made to fit a $\frac{3}{4}$-inch bead, as shown in No. 1, fig. 1179, but special hooks can be made for special mouldings, as in Nos. 2 and 3. The last shows a combined cornice and picture-mould, suitable for low dining-rooms where some of the pictures are usually of large size.

Plate-rails are sometimes fixed around the rooms, and may have a bed-moulding shaped to serve as a picture-mould, after the manner of No. 3, fig. 1179. No. 4 shows a plate-rail on the top of oak-panelling, designed by the writer for the billiard-room of a Suffolk house. The special feature is the groove along the top to receive the edges of plates, photographs, &c.

Pews.—Pews in churches and chapels are usually constructed to give a seat-length of 18 or 20 inches for each person, and the backs are from 30 to 36 inches from centre to centre. A convenient spacing is 33 inches, as shown in fig. 1180, which gives three different designs for pews. In the first the book-shelf is supported on brackets, these being often spaced to show the number of "sittings" in the pew; a drawer for books is formed under the seat, and is opened from the pew behind; the space from the seat to the floor is filled with vertical match-boarding. In the second example a rack takes the

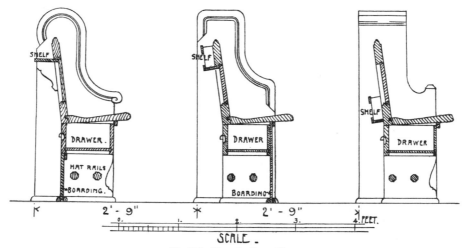

Fig. 1180.—Three Sections of Pews

place of the simple shelf, and the boarding under the seat is placed in a different position to allow more room for the feet and to allow hats to be placed in sight of the owners. In the third example the book-rack is at a lower level, and the boarding under the seat is omitted. The seats are hollowed on the top, the front edges being rounded, and are $1\frac{1}{2}$ inch thick, and supported on shaped standards about $1\frac{1}{2}$ inch thick and from 3 to 4 feet apart; the ends of the seats are housed into the "pew-ends". The backs are of $1\frac{3}{4}$- or 2-inch framing, with flush panels of narrow V-jointed or beaded match-boards. The pew-ends are usually $1\frac{3}{4}$ inch or more in thickness, and are screwed to the flooring and to the seats and backs. In many cases, the backs extend down to the floor. An example of pews for the gallery of a chapel will be found in Plate XLIV.

Plate CVIII shows the lower part of one bay of the perpendicular screen in Southwold Church. The screen is of oak, richly moulded and carved, the tracery being carved out of the solid, and the framing being secured with oak trenails.

CHAPTER VI

GATES AND FENCES

A gate may be regarded as a movable portion of a fence or enclosure, and often takes the form of a ledged, framed, or panelled door with a square, chamfered, or moulded capping to prevent the entrance of rain into the upper edge. But in many cases such a gate would be objectionable on account of its weight and costliness; and a rectangular frame of wood,

sparred or barred in such a manner as to prevent the passage of animals, and to give the necessary rigidity, is therefore substituted.

In ordinary field-gates the width of the opening between the posts, to which the gate is hinged, is about 9 feet. The height of the gate is often regulated by the height of the fence of which it forms a part. The vertical side-pieces of the framing are called stiles; that to which the hinges are attached being called the hanging-stile, and that to which the fastening is attached the falling-stile. The horizontal members are called rails.

Such a frame suspended by one of its shorter sides would not maintain the rectangular form; it would become rhomboidal by the falling down of the other sides by their own weight. To enable it to maintain the rectangular form it is necessary to add an angle brace, which may be applied either as a tie or a strut, as the material used is iron or wood. Let the diagram (No. 1, fig. 1181) represent—a the hanging-stile, b the top-rail, and c the brace or strut of a gate, all firmly united. This is evidently a simple truss, like the jib of a crane; and if a weight w be hung to its outer end, the rail b will obviously be in a state of tension, and the brace c in a state of compression; that is, b is a tie and c a strut. Again, let a (No. 2) be the hanging-stile, b the lower rail, and c the brace, and it is now obvious that c is a tie and b a strut. Therefore, keeping in mind that iron should be used as a tie and

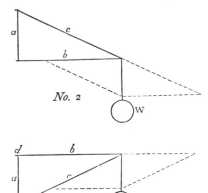

Fig. 1181.—Stresses in Gate

wood as a strut, when the brace is placed as in No. 1 it should be of timber, and when as in No. 2 it should be of iron. It may be objected that, if this rule were adhered to, it would not be possible to construct a timber gate, as it requires both ties and struts. But the straps by which the hinges are attached to the frame are generally so long as to embrace a considerable portion of the length of the rail, and therefore render the timber rail b (No. 1) competent as a tie; whereas if timber be used for the brace c, in No. 2, it cannot be so easily reinforced, as both ends must be firmly secured.

Fig. 1182 is the ordinary field-gate, constructed on these principles. The top rail a becomes a tie, and is secured to the hanging-post by the strap of the upper hinge embracing it, and being bolted through it. The elementary frame is thus rendered perfectly rigid, and the addition of the front or falling-post b, and of the bars e, f, g, h, completes the fence. The strut should be attached to the top bar, at such a distance from the end of the latter as to afford a perfect abutment; in ordinary cases, 10 inches or 1 foot is sufficient. The hanging-stile is $4\frac{1}{2}$ inches square in section, and the falling-stile 3 inches square. The top

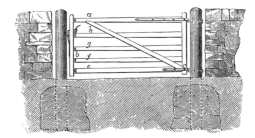

Fig. 1182.—Simple Field-gate

rail a is $4\frac{1}{2}$ inches square at the hanging-stile, and 3 inches by $4\frac{1}{2}$ at the falling-stile. It is tenoned into the stiles. The diagonal bar m is $4\frac{1}{2}$ inches deep, and 2 inches thick; tapering to the upper end for the sake of lightness. It is tenoned into the hanging-stile, and notched into the top rail. The other rails e, f, g, h are $4\frac{1}{2}$ inches deep at the hanging-stile, and taper to $3\frac{1}{2}$ inches deep at the falling-stile. They are $1\frac{1}{2}$ inch thick, and are tenoned into the stiles with barefaced tenons. The upper hinge has its straps prolonged to embrace a considerable portion of the top rail. The lower hinge does not require this prolongation of the straps, as the force upon it is a thrusting, and not a drawing force.

As the bottom rail is so much thinner than the hanging-stile, a small piece of wood, of the depth of the rail, is generally added at one end as a rest for the strap of the hinge.

The tenons of the rails are secured in the mortises by pins, and the diagonal is securely nailed to the rails at its intersection with them. Before putting the parts together, the tenons and the intersecting parts of the rails and diagonals should be coated with white-lead in oil. The great destroyer of the gate is rain, which, falling on the thin top bar, as usually constructed, soaks into the joints and induces rot. The wide top rail in the gate described affords protection against this; and the only parts exposed to it are the intersections of the diagonals and rails; but by giving the upper edges of these a slight bevel,

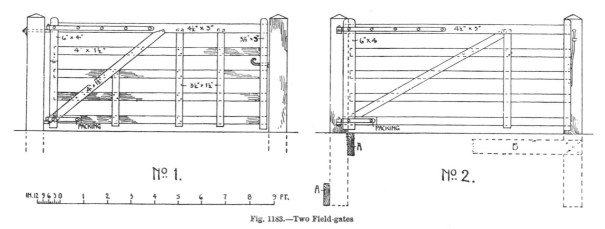

Fig. 1183.—Two Field-gates

so as to throw the water from the joint, the risk of injury from this cause is reduced. The top rail should be saddle-backed or rounded on the top. Sometimes vertical bars or "pales", about $3\frac{1}{2}$ inches by $\frac{3}{4}$ inch, are nailed to the framing on the face of the gate; they render the gates more difficult to climb, but form a number of joints which may admit rain and thus hasten the decay of the wood.

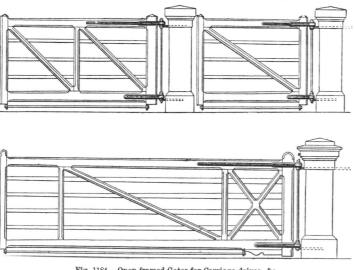

Fig. 1184.—Open-framed Gates for Carriage-drives, &c.

Other forms of field-gate are shown in fig. 1183. In No. 1 the hanging-stile is out of 6 × 4-inch stuff, the top rail out of $4\frac{1}{2}$ × 3-inch, falling-stile, $3\frac{1}{2}$ × 3-inch, the bottom, intermediate, and diagonal rails are 4 × $1\frac{1}{2}$ inches, and the verticals, $3\frac{1}{2}$ × $1\frac{1}{2}$ inches. Sometimes the top rail is tapered from $4\frac{1}{2}$ × 3 at the hanging-stile to $3\frac{1}{2}$ × $2\frac{1}{2}$ at the falling-stile, and the falling-stile is made $3\frac{1}{2}$ × $2\frac{1}{2}$, the other members being also reduced in size. In No. 2 the same scantlings may be used. The tenons in some cases do not run through the stiles. The different members are secured with wedges, trenails, and iron nails. The verticals are secured to the top rail with barefaced tenons. The foot of the diagonal is notched or tenoned into the hanging-stile. The posts may be simply fixed in holes in the ground by packing around with stones or suitable earth well consolidated, or may rest on a bed of concrete 6 inches thick, and be surrounded with concrete up to the surface of the ground. Sometimes the stability of the hanging-post is increased by means of two pieces of 11 × 3-inch planking about 3 feet long, spiked to the outer and inner sides, as shown at A A in No. 2, or the two posts are braced together by one or two deals or battens, B, spiked to the

PLATE CVI

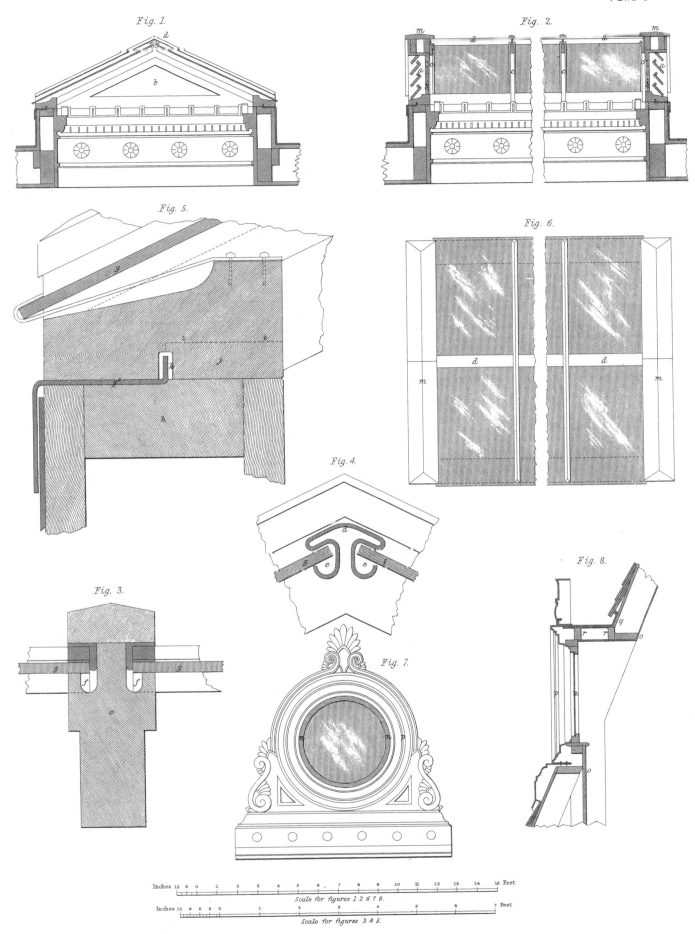

Fig. 1.

Fig. 2.

Fig. 5.

Fig. 6.

Fig. 4.

Fig. 3.

Fig. 7.

Fig. 8.

Inches 12 6 0   1   2   3   4   5   6   7   8   9   10   11   12   13   14   15 Feet

Scale for figures 1 2 6 7 8.

Inches 12 9 6 3 0   1   2   3   4   5   6   7 Feet

Scale for figures 3 4 5.

SKYLIGHT AND DORMER

PLATE CVII

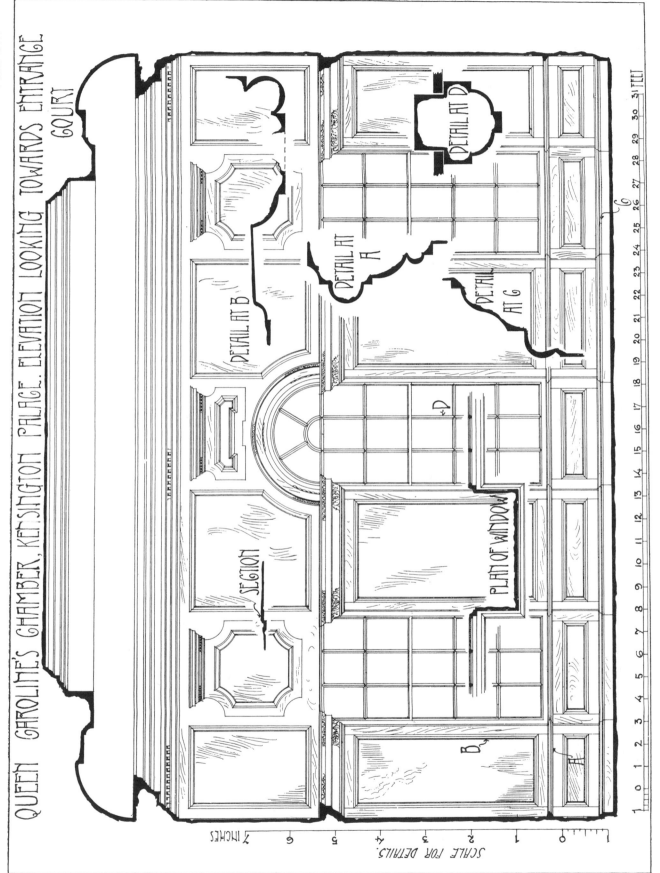

QUEEN CAROLINE'S CHAMBER, KENSINGTON PALACE. ELEVATION LOOKING TOWARDS ENTRANCE

COURT

DETAIL AT D

DETAIL AT A

DETAIL AT G

DETAIL AT B

SECTION

PLAN OF WINDOW

SCALE FOR DETAILS.

7 INCHES

PANELLING, &C., QUEEN CAROLINE'S CHAMBER, KENSINGTON PALACE

PLATE CVIII

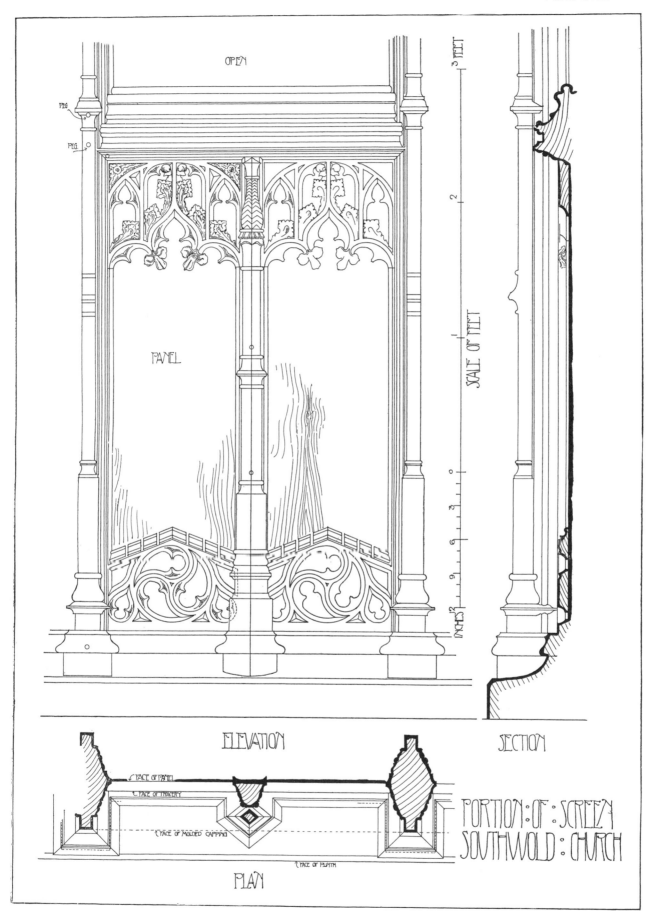

OPEN

PEG.

PEG.

PANEL

SCALE OF FEET

3 FEET

2

1

0

INCHES

ELEVATION

SECTION

℄ FACE OF PANEL
℄ FACE OF TRACERY
℄ FACE OF MOLDED CAPPING
℄ FACE OF PLINTH

PLAN

PORTION · OF · SCREEN
SOUTHWOLD · CHURCH

PORTION OF CHANCEL SCREEN, SOUTHWOLD CHURCH

sides of the posts. All the wood below ground, and to a height of not less than 6 inches above ground, ought to be creosoted, tarred, or charred.

Two examples of more elaborate open-framed gates, suitable for carriage-drives, &c., are given in fig. 1184, one-half of the gateway being shown in each case. No. 1 has a pair of gates in the centre for car-riages, and two small gates at the sides for foot-passengers. Fig. 1185 shows a gate de-signed by the writer for a house near Woking, and car-ried out in teak; the lower panels are filled with $\frac{3}{4}$-inch tongued-and-grooved V-jointed boarding flush with the framing on the face of the gate. The principal members are 3 inches thick, and the diagonal $2\frac{1}{4}$ inches thick and halved to the muntins.

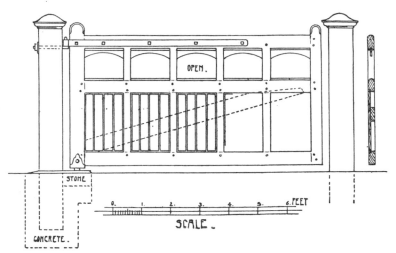

Fig. 1185.—Gate for Carriage-drive

All the gates which have been described have wood struts. Sometimes these struts are omitted, and wrought-iron straps are bolted to the wood framing, one on each side, extending from the head of the hanging-stile to the foot of the falling-stile, to serve as ties, as explained in No. 2, fig. 1181; but such gates often twist to a considerable extent, and this method of construction cannot therefore be recommended.

Two panelled gates are given in fig. 1186. In No. 1 diagonal struts ought to be inserted behind the panelling to keep the framework rigid. In No. 2 two small rollers are

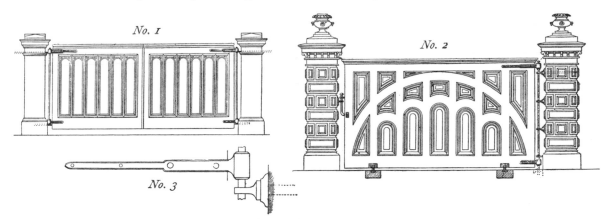

Fig. 1186.—Panelled Gates

inserted in the bottom rail, which run on iron guides, laid on stone or concrete sleepers fixed in the ground. The hinge-pin is continuous between the top and bottom hinges, and serves merely as an axis on which the gate rotates, the whole of the weight being sustained by the rollers. It has sufficient play to allow the gate to rise as it opens. No. 3 is the strap of the top hinge, and the same construction of strap is applicable to all the previous examples, where the object is to extend the hold of the strap on the top rail. The length of the strap may with advantage be increased to rather more than half the width of the gate. The strap of the bottom hinge may be very short, although, for the sake of uniformity, it is often made the same length as the upper strap.

Fig. 1187 shows two pairs of gates with cast-iron panels. Details are given of the

head and heel of the hanging-stile of No. 1, and of the upper part of No. 2. The gates in
No. 1 are hung with strap hinges at the top and the heels of the hanging-stiles rest on

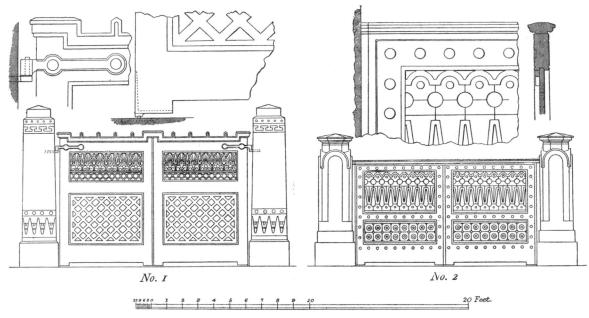

Fig. 1187.—Gates with Cast-iron Panels

pivots at the bottom secured into stones. In No. 2 the gates are hung with 7-inch edge
hinges, to stiles of the same thickness as the framing of the gates, which are fixed to the
pillar by iron batts.

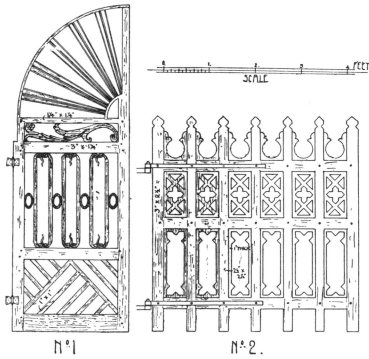

Fig. 1188.—Two pairs of Swedish Gates

The examples in fig. 1188
were measured by the writer in
Sweden. No. 1 shows one-half
of a pair of very light gates in an
archway at Hudiksvall, and No. 2
shows one-half of a pair of low
gates near Gefle, with fretwork
panels. The scantlings are figured
on the drawings.

The entrance-gates to church-
yards are often placed in a framework supporting a roof, and are then
known as lych-gates. An interesting example is given in fig. 1189.

Gate-fittings.—Hinges of the
kinds shown in fig. 1139 are commonly used for gates, but other
methods of hanging may be adopted
as in figs. 1186 and 1187. Another
form of hinge (fig. 1190) is specially
adapted for rising ground; No. 1
shows Collinge's double-strap upper
hinge, and No. 2 the bottom self-closing hinge. As the gate is opened, the lower hinge
takes a different bearing, which throws it out of plumb, and causes the gate to rise as it
opens. These hinges are made with backplates, as shown, for screwing or bolting to wood
posts, or with prongs for fixing in stone or brick.

Fig. 1191 contains illustrations of some other fittings for gates; No. 1 is a self-acting rising and falling stop for bedding in concrete under the meeting stiles of a pair of gates; Nos. 2 and 3 are weighted catches for holding gates open, the first being for fixing in stone or concrete, and the second for screwing to wood; No. 4 is a simple latch fixed in a mortise

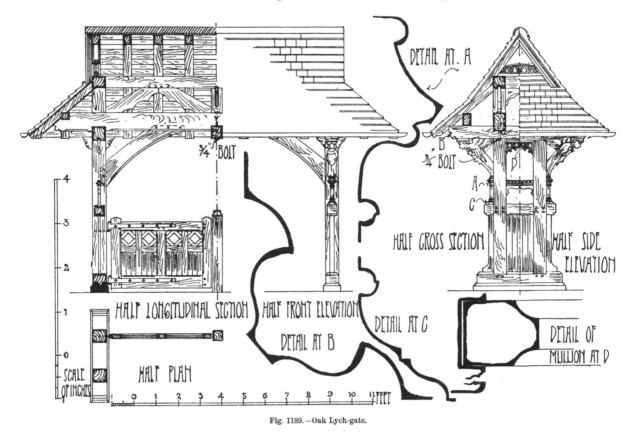

Fig. 1189.—Oak Lych-gate.

in the falling-stile; Nos. 5, 6, and 7 are strikers to receive the latches of gates which do not open through, the last having lugs for fixing in stone; Nos. 8 and 9 are strikers for gates which open both ways; and No. 10 is a hinged hasp and staple for a padlock. Latches are now made in endless variety, many having ring handles, and others being of similar type to the mortise latches of doors with iron or gun-metal knobs.

Fencing.—The simplest form of wood fence is that known as "post and rail". The

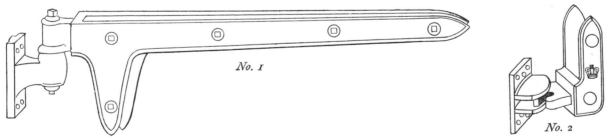

Fig. 1190.—Collinge's Top Hinge and Self-closing Bottom Hinge

posts are fixed from 6 to 9 feet apart, and may be of oak, pitch-pine, or larch. The ends of the posts may be sharpened and tarred, and driven into the ground about a foot, or to a greater depth if the ground is soft. The rails may be nailed to the posts or passed through mortises. Two varieties of post-and-rail fencing are shown in fig. 1192. In No. 1 the posts are $5\frac{1}{2} \times 4$ inches, and 6 feet from centre to centre; the tops are bird's-mouthed to receive the 4×4-inch top rails, which are secured to the posts by hoop-iron

straps nailed on; the lower rails are $4\frac{1}{2} \times 2$ inches, and 13 feet long, and are passed through mortises in the posts, the ends being scarfed as shown and nailed.   In No. 2 the main posts are of oak, 6×6 inches, and 12 feet from centre to centre, the butt ends being tarred and

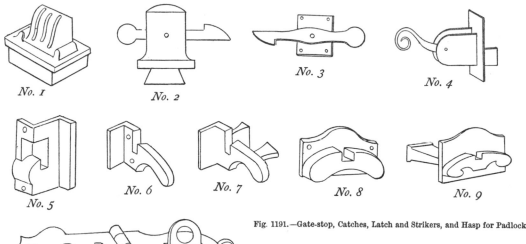

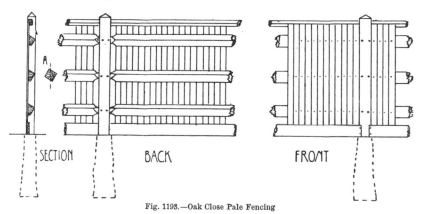

Fig. 1191.—Gate-stop, Catches, Latch and Strikers, and Hasp for Padlock

bedded in the ground.   An intermediate oak post, about 4 by 2 inches, is pointed at the end and driven into the ground, and nailed to the rails, which may be of oak or red deal fixed as in No. 1.

Open pale fencing consists of posts and rails with vertical pales about $3\frac{1}{2} \times \frac{3}{4}$ inches

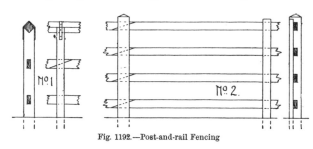

Fig. 1192.—Post-and-rail Fencing

nailed to the rails. For fences up to $4\frac{1}{2}$ feet in height, only two rails are required, but for greater heights, up to 6 feet, an intermediate rail ought to be introduced.   Larch is often used for these fences, either sawn or wrought, but oak is more durable, and when this is used the pales may be cleft.   Close pale fencing (fig. 1193) is usually of oak, the posts being about $5 \times 4\frac{1}{2}$ inches for fences $4\frac{1}{2}$ feet high, and $6 \times 4\frac{1}{2}$ inches for fences from 5 to 6 feet high.   The posts are bedded in the ground to a depth of $2\frac{1}{2}$ or 3 feet, and are fixed 9 feet from centre to centre.   The rails are cut out of stuff of the same scantling as the posts, as shown at A, and are tenoned into the posts, two being used for fences up to $4\frac{1}{2}$ feet high, and three for higher fences.   The pales are cleft, $3\frac{1}{2}$ to $4\frac{1}{2}$ inches wide, and are often feather-edged; they are fixed to overlap about $\frac{3}{4}$ inch, and are secured to the rails with galvanized

Fig. 1193.—Oak Close Pale Fencing

steel or iron nails.   Gravel-boards from 6 to 11 inches wide and from $\frac{3}{4}$ to $1\frac{1}{4}$ inch thick, are often fixed along the bottom of these fences, and the top of the fence is sometimes protected by a coping, nailed to a rectangular top-rail.   Somewhat similar fencing is made of sawn larch, coated with Stockholm tar.

Section XIII.—SHOP MANAGEMENT

BY

A. C. REMNANT

CHAPTER I

SHOP ARRANGEMENT, &c.

The important question of the arrangement of shops and machinery appears to have been almost entirely neglected in many shops, particularly in the older-established businesses. This is owing, no doubt, to the fact that at their inception hand labour was nearly the only kind in existence, and labour-saving devices, when introduced, had to be adapted to the premises at command instead of *vice versâ*; the result is that the arrangement, as regards economy in working, leaves much to be desired. Premises with any pretension to size or repleteness will, amongst other accommodation, comprise the following:—Machine and Joiners' Shops, Engine-Room and Boiler-House, Mill-Office, Latrines, Stabling, Shedding, and racks for the storage of timber.

It is very important in the planning of shops, and of machine-shops in particular, that the machines should be so placed as to be capable of being supplied with material in a ready manner, and also that a certain amount of space should be allowed contiguous to each machine in order to allow the temporary placing of the rough and finished material.

The size of shops will be regulated by the amount of present or prospective work, the size and situation of the site, financial and other causes. For a small business housing accommodation for half a dozen (or fewer) machines may be all that is necessary, whereas for larger works the machines in general use may be duplicated or triplicated, thus requiring a considerably larger area for their accommodation; where machines are in duplicate, it is better to arrange those of each kind in close proximity to each other.

For ordinary mills a convenient width is about 50 feet. This will be found sufficient to allow for efficient side lighting, and for a passage or tramway down the centre with turn-tables as required, and also for lines of machines on either side. Overhead joiners' shops can be arranged on the gallery principle for one or more floors as required, with a central well about 10 feet wide, in which a hoist or lift can be placed to convey material from the machine-shop to the upper floors. Sliding-doors are recommended in preference to swing-doors, and the internal sills of the windows should be weathered so as to afford as little lodgment for dust as possible. A convenient height for shops is from 12 to 15 feet.

Tramways and trolleys will be found to be almost indispensable adjuncts; many of the expenses of a wood-working manufactory can be charged to the handling of material, and the acquisition of these means of transport will effect a considerable saving in this depart-ment. The trucks can be made of stout timber with framed ledged bottoms and swivel castors so as to move readily in any direction, or they may be made of light angle-irons with similar castors. If every machine is provided with at least one trolley, half the cost

369

of handling and nearly the whole cost of carrying will be saved. The trolley system should be carried out in its entirety, for if the trucks are limited in number, annoyance is caused and time lost by men waiting for each other or searching for an available truck, and when one is found (possibly in use), further time is lost in wrangling over its possession.

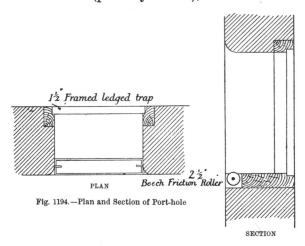

1½" Framed ledged trap

PLAN

2½"
Beech Friction Roller

SECTION

Fig. 1194.—Plan and Section of Port-hole

Where shops are arranged on upper floors, or the gallery system is employed, a hoist or lift is an absolute necessity. Trolleys can be easily wheeled on to it, raised to any floor, and their contents deposited at convenient points, thus effecting a great saving in manual labour. A wooden platform or cage, with a wire rope and winding drum driven by bands and a tangent wheel, is a cheap and simple plan for such a hoist, or reference to the catalogues of any of the well-known lift engineers will readily suggest appropriate plant for this purpose.

With a machine-shop of a width of 50 feet, the shafting should be arranged in transverse lines rather than longitudinally, so that the shafts can be worked at different speeds when required. The last shaft, or the one farthest from the engine, can, for example, be driven at a higher speed than the others to serve machines on upper floors as and when necessary. Wherever possible it is advisable to have the shafting and gearing underground in trunks, with manholes at intervals

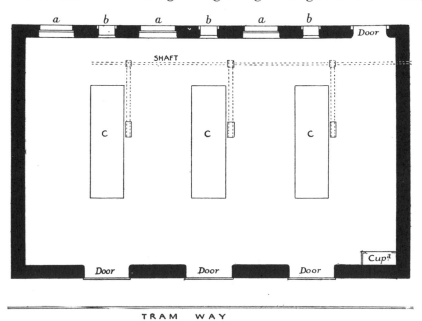

Fig. 1195.—Plan of small Planing-Mill
a a a, Windows. b b b, Port-holes. c c c, Planing-machines.

for access. The adoption of this method lessens the danger of employees being caught in the belts, and, if air-tight covers are provided to the inspection-holes, dust and dirt cannot reach the bearings to the same extent as in the case of overhead and exposed shafting. The most usual and best sizes for shafting are $2\frac{1}{2}$ and 3 inches; in fact, these may be taken as the minimum and maximum sizes respectively for wood-working machinery. Pulleys should always be carefully balanced and turned true, and where overhead shafting is in vogue, the hanging-irons should have their bearings pivoted and adjustable vertically, in order to level up and line the shafting when required.

Where, as in the case of planing-mills, the material is merely dressed or matched and jointed, a narrower shop can be used with advantage, the shafting being arranged longitudinally, and the machines placed transversely, with doorways of sufficient width opposite the machines at one end for the introduction of raw material and small openings in the wall opposite the other end to pass prepared material out. The speedy removal of

the converted material is an important point. These small openings or portholes should be fitted with free-running rollers to reduce friction, as shown in fig. 1194. The arrangement of a planing-mill such as last described is fully shown in fig. 1195.

New shops should be constructed in a fireproof manner, with walls and floors of sufficient strength to resist the jar and vibration of machinery running at high speeds; ground-floors in particular should be exceptionally strong on account of the heavy machinery and possible storage of timber. The first is not so serious a factor as the second, as the machines are generally carried by separate foundations to keep them level and rigid, whereas the storage of large piles of timber may cause depressions in the floor and a consequent list of the machines. It will be found of great convenience, when new premises are being built, if each floor has what may be termed a "machine line" or base, from which the machines, shafting, &c., may be set out independently of each other and yet with accuracy; such a line obviates the setting out of new machinery by parallel lines with the existing machines, and the arrangement of new machines becomes a very easy matter, as lines parallel to or at right angles to the base line are easily laid out and shafts can be set true by measurement. This line should be made through the building both ways and scribed or marked on the floors, walls, or ceilings. When setting out machines, care should be taken to set them level, with the shafts and spindles parallel to the line-shaft, and to see that there is always truck-room between and around the machines. The floor-space saved by piling stuff on trucks will more than counterbalance that lost in passages.

In the arrangement of shops, sheds, stores, &c., the proximity of the raw material to the centre of conversion must be considered, so that excessive cost is not entailed in handling the material; this cost can be further reduced if a system of tramways from the shop to and past the racks of timber is adopted. It will be found very handy to have racks for storing fillets and timbers of small scantling; such a method further allows of thorough natural seasoning. If space can be spared, it is very useful to have a covered shed—commonly known as "the birdcage"—with open lattice sides, in which prepared joinery can be further seasoned before being sent to its destination. Mills which have any pretension to size and completeness should be fitted with a separate room or compartment in which saws, moulding and plane irons, &c., can be ground and stored.

It will also be found advantageous to have an office in the mill where drawings of work in progress, and books and papers relating to the cost of mill-labour can be kept, and from which the foreman can exercise a thorough supervision of men and machinery, and in which, also, the setting-out can be done. The upper part of the sides of such an office should be glazed, and a small hatch for enquiries to be made through should be formed so as to obviate the possibility of any workman prying when talking to the foreman or mill-clerk. This office should be fitted with tables on which to set out the rods, and under the tables can be provided nests of drawers in which to keep drawings. Racks for books and other fitments can be provided as circumstances require.

Sanitary arrangements for the men should be provided in a ratio proportionate to the number employed. The regulations of different local authorities range from one closet for twenty men to one for forty. Urinals must also be provided. The conveniences should be arranged as nearly central as possible, so that little time is lost in approaching them. Trough-closets or latrines will be found most suitable, and should be automatically flushed at stated periods with a flushing-tank. The seats may be about 4 inches wide and rounded on the inner and outer edges; if separate basins are provided, the seats may be simple strips of wood (with rounded edges) fixed to the sides of the basins, the pottery being exposed to view at the front and back. Divisions about 6 feet high and 6 inches from the floor should be fixed between the closets, with "decency doors" in front, hung on strap hinges and fitted with bolts. Urinals can be fitted with a glazed-ware channel, 6 or 9 inches

in diameter, bedded on brick-work, and flushed every half-hour by an automatic flushing-tank.   Latrines should be lime-whited at frequent intervals.

The best modes of natural lighting are from the sides and roofs of mills, and it will be found a profitable investment if the windows are protected by wire-netting on the inside, as short ends of stuff have a knack of hurtling through space, and broken squares of glass are a frequent result.   In a joiners' shop the windows should be fixed so that the sill is, say, 9 or 12 inches above the top of the bench opposite, as a mechanic engrossed in his work will sometimes run the material he is working on, past the end of his bench and through any glass opposite it.   In some establishments the man working opposite the window is responsible for breakages, but the collection of the amount of damage or stopping it out of wages on pay-day is only productive of friction and ill-feeling.

Regarding artificial lighting, it will be necessary to consider the means at disposal. In many of the large mills on the Baltic electric light is used, and where large volumes of light are required there is not the least doubt of its excellence.   It will be found most profitable for mills to have their own installation, the laying down of an engine with a little extra horse-power beyond that required for driving the machines being sufficient to work a dynamo; if desired, a separate engine of the requisite power can be utilized for this purpose. Where, on account of cost or other reasons, electric light is precluded from use, recourse must be had to gas, or (for the yards) to illuminants of the Wells-lamp type.   Where gas is used the burners should have wire shades fitted to obviate, as far as possible, contact with shavings.   Swing T-pendants should be fitted to benches and machines, so that when not required for use they can be swung up and placed out of the way.

The general causes of fire in wood-working shops may be attributed to smoking, matches, stoves, sparks from a furnace, heated bearings, lightning, and incendiarism. Smoking should not be allowed under any circumstances on the premises; matches are not dangerous if carefully used; stoves, where required for heating glue or other purposes, can be rendered safer by setting them on an elevated iron platform, but in the present days of glue-heating by steam from the boiler, all risk of fire in this direction should be practically annihilated.   In some shops gas ring-burners are used for heating glue, and these can be conveniently arranged in niches in the walls.   By such means it is comparatively easy to allocate glue-pots to certain sections of the men.   Danger from furnace-sparks implies negligence in the arrangement or construction of the furnace, or in its care.   As to bearings becoming heated, these should be so made that there is no wood about them, and no accumulation of shavings, or oil and sawdust.   As it is a difficult matter to fire a shop that is free from an accumulation of shavings, it is imperative that shops and floors should be kept clean.   Further, on every floor and in convenient positions should be racks with pails of water, and care should be taken that these pails are always filled, and that their contents are not used for other purposes, such as washing, filling glue-pots, &c.   Each pail should be marked FIRE, and the water can be kept pure by the addition of a few drops of carbolic acid.   In addition to buckets, hydrants should be fixed at suitable positions in the buildings; they should be enclosed in cases with doors fitted with glass panels, which can be readily broken in case of necessity to gain access to the fittings.   The hose should be rolled up and lightly secured with thread, and should have a bayonet-clip connection so as to be immediately adjusted to the hydrant.   It is also advisable in large establishments to form a shop fire-brigade, and to arrange for weekly drills. Members of such a brigade would, by knowing the general plan and arrangement of the premises, be capable of rendering more efficient aid than the regular firemen.   A certain amount of interest taken in such a brigade by the master would ensure a similar interest on the part of the men.

Amongst automatic appliances for the extinction of fire, that known as the Grinnell Automatic Sprinkler and Fire-Alarm holds a leading position.   The installation consists of a series of pipes attached to the ceilings of all rooms in the shops.   These pipes are charged

with water supplied under pressure from an elevated tank, pumps, or public water-supply. To the various branch-pipes the sprinkler heads are attached at distances not exceeding 10 feet apart. The sprinkler head consists of a valve held in a closed position by metal supports secured with solder fusible at a low temperature (155° Fahr.). In the case of a fire occurring in any part of the factory, the heat immediately fuses the solder of the nearest sprinkler. This automatically releases the valve, and the water descends in a dense shower, covering an area comprised in a circle of about 15 feet diameter. Simultaneously with the sprinkler opening, a gong is automatically sounded, and this being effected by the action of the water as it passes through the pipes, does not cease until the watchman, or whoever may be on the premises, finds it advisable to turn off the water. No human agency is required to operate the Grinnell sprinkler, which comes into action automatically under the influence of heat. In cases where the pipes are likely to be subject to frost, the dry-pipe system is used. In this case the pipes are filled with air, which immediately rushes out on one of the valves opening, giving place to the water which immediately follows.

To isolate adjoining shops armoured fireproof doors are now generally recommended in place of the old-fashioned iron doors, so liable to buckle and warp when subjected to any great heat. These doors are constructed of two, three, or four thicknesses (according to the area of the opening) of well-seasoned pine boards, covered on both sides and on the edges in a special manner with tinned steel plates. By excluding the atmosphere from the wood, combustion is prevented, an exposure of several hours to the flames only resulting in the timber becoming carbonized to the depth of a fraction of an inch.

Notwithstanding the fact that the removal of shavings, sawdust, and other débris arising from a wood-working establishment is one of the most important questions in connection with its economical working, it is remarkable that until within comparatively recent years only very few mills and steam joinery works possessed any special provision for the purpose. Oftentimes it was left to a weekly clean-up by the shop lads, the refuse being put into sacks or taken to the stoke-hole for firing, or the sawdust and shavings were put into separate sacks and sold—for in some districts a market can be found for them, horsekeepers preferring machine shavings for littering stables. Independently of anything else it is most essential as a security from fire that mills should be thoroughly cleaned up at least once a week. The flanges of girders, brackets, &c., likely to prove resting-places for fine dust should be particularly attended to, for should by any chance any become dislodged, say by a draught, and get in contact with a gas flame, a serious explosion would probably result; the fine dust also enters the respiratory organs of the workers, with the worst possible effects. To minimize these risks the installation of a pneumatic system for clearing the mills of wood-refuse is one of the best safeguards. The installation consists of a blower or exhaust fan, and the suction-pipes, which connect the fan with the machines from which the refuse has to be removed. The shavings are sucked up into these pipes and carried to the fan, and thence by delivery-pipes to the collecting-room. In years gone by difficulties were experienced in separating the very fine dust from the air, and the dust escaped with the current of air through the openings in the collecting-rooms, covering the surrounding buildings, and blocking up the gutters and rain-water pipes, and causing an annoyance to residents in the neighbourhood. Now, however, thanks to the use of separators, which have taken the place of the collecting chambers, these difficulties have been overcome. The separator is made of iron plate, and is conical in shape; the delivery-pipes enter the upper part, and by means of a special arrangement the current of air, saturated with dust, receives a spiral rotary motion, and the shavings and dust fall down by reason of their own weight, whilst the air escapes, entirely free from dust, through an aperture at the top of the separator; as the air-blast escapes thus easily, the fan is relieved of all back pressure, and hence requires less power to drive. Readers who desire more information will be well repaid by studying the catalogues of such firms as A. Ransome & Co., or Kirchner & Co.

Care should be taken that waste wood is not thrown into the exhaust pipes. Such hard matter striking against the fan wheel, rotating at an enormous speed, may break it.

Accidents in wood-working establishments generally occur from carelessness on the part of operators, or procrastination in repairing or correcting an irregularity or risk in a machine. A man who has neglected to study the sources of danger to which he is exposed in his vocation, exposes himself to risk that may cost him a limb or his life at any moment. Familiarity, no doubt, breeds contempt, and one often sees machinists running the greatest risk by their dangerous proximity to machinery in motion. Want of thought also accounts for many accidents. A spanner may drop under a saw-bench; the operator dives his hand underneath, groping for it without thinking of danger, and his hand gets in contact with the saw and is possibly amputated. Again, a man at a planing-machine, in guiding the stuff with his hand, overlooks the revolutions of the iron, and allows his hand to follow the course of the material over the cutters, with a resulting loss of two or three fingers, if nothing worse. These are facts which cannot be disputed, for there are very few operators who are not maimed in a more or less degree. No doubt these accidents are due to the, in many cases, small size of a machine or its apparent harmlessness, and the absence of noise also seems an additional temptation. Absent-minded men, or those given to day-dreaming, are sure to be injured, and a manager by carefully observing the habits of his staff can select men who are liable to run little risk. Should a man be found daydreaming, or his mind wandering from his work, for his own sake let that man be removed from a dangerous machine and given one with less risk.

Perhaps the most dangerous machines are saws (both band and circular), planing and moulding machines. With these the hands are of necessity continually exposed to injury, and nothing but continual attention and care will prevent accidents. Many accidents are due to the hands being jerked or dragged on to the saw or planing-machine. When cutting or planing short pieces, the machinist should use a stick in his right hand for pushing the pieces, placing his left hand to keep them against the fence. There are, of course, safety guards, shields, and checks in the market to limit the dangers from throwing back, but instead of being used, these guards are often discarded by the machinist, and may be found hanging on a wall near the machine. Any safety device that impedes or increases labour is bound to be disregarded by men who apparently prefer to run into risks rather than prevent them. Vigilant Board of Trade Inspectors, and the provisions of the Employers' Liability Act, have tended to increase the employers' interest in the safety of their employees, but employers cannot prevent accidents if the workmen are careless.

CHAPTER II

MANAGEMENT OF MEN—TRADE-UNION RULES, &c.

In these days of combination on the part of both masters and men, umbrage is often taken for the most trivial causes, and lock-outs and strikes are a frequent consequence. It may safely be recognized that the days of bullying and bluster on the part of a master are gone, and that a spirit of conciliation and fairness to employees is more frequently shown by employers than used to be the case in the past. Let each meet as man to man with a knowledge that the one pays for the work of the other, and that a certain amount of trust must be placed in each other; let the master (as some do) take an interest in his men and their welfare (this can be done without any loss of dignity), and one may venture to say that the tone of labour throughout the country will be raised.

One thing is most essential in the management of men. Masters and their foremen

should always keep their position; they should not dispute in the shops or on the building, nor should they drink together at a public-house. Nothing lowers a master so much in the eyes of men as this habit. There is, however, no reason why an employer should not associate with his men in their pastimes. The formation of athletic or benefit clubs, with the principal as president, invariably tends to promote good feeling between all concerned, and if interest is shown by "the boss", it is always appreciated by the men.

When considered from the point of mutual help, the advantages of trades-unions are manifest. Some unions state that they exist for the purpose of providing funds for the support of their members when travelling in search of work, and to afford them assistance when illness or old age creates inability to labour. They also provide for the burial of members and their wives. But objects such as the above, although important to workmen, who are liable to be reduced to want by many accidents that cannot be forestalled, do not as a rule comprise all the aims of a union, otherwise it would be merely a provident society. The main object of a union is to organize and systematize combination—that power which is slowly making its strength apparent throughout the land, and, in some cases, reducing employers of labour almost to a condition of impotence. It is a power also which is raising workmen from the dependent state in which they existed for many years, and, by the inauguration of strikes, a number of men can, if their funds permit, force the hands of employers in many instances, and compel agreement to their demands. The power (financial and moral) which employers possess, and a sometimes unscrupulous abuse of it, rendered combination on the part of workmen a necessity to protect their interest.

What, then, does this combination imply on the part of the trade-unionist? In the 1899 rules of the Amalgamated Society of Carpenters and Joiners, there are upwards of eighty fines and penalties which the member may incur, ranging from 3d. for "neglecting to attend a summoned meeting", to £2 for a "third offence of disorderly conduct or causing a quarrel at the Society's meeting". Some of the fines border on the ridiculous; for instance, "If one member upbraids another for receiving benefits", he is mulcted of 2s. 6d.; "If a member censures another concerning the Society's business", he is eased of 1s., whilst if a member in his early experiences of connubial bliss fails to give notice of his marriage, his oversight costs him 2s. 6d. It is somewhat regrettable that one rule, which was in force in 1893, should have been eliminated, namely,—"If a man boasted of his independence to an employer", he was fined 2s. 6d. for each offence. This rule no doubt curbed the tongues of many men, and the mechanic who lent me his 1893 rules has, I notice, put a pencil note "*Good*" against this rule.

In addition to conforming to these rules, a trade-unionist must resolve to set aside the immediate advantages of himself, his neighbours, or his district, for the general benefit of the Society to which he belongs. He must avoid petty squabbles and divisions, and disregard local and personal interests. He must be obedient to the local committee, and at their command must leave constant work, a good master, and possibly 40s. or 50s. a week, in order to loaf about the streets on a pittance of 10s. or 12s. a week (strike pay).

The rate of wages per hour and the number of hours worked per week vary throughout the country, but, generally speaking, in large towns the rate is higher and the number of hours less than in the small towns and villages. The following table gives the extremes at the beginning of 1902, from the returns prepared by the Labour Department of the Board of Trade:—

ENGLAND AND WALES.				Working hours per week.		Rate per hour.
Northern Counties and Cleveland ...	...	...	...	50 to 55	......	$5\frac{3}{4}d.$[1] to 10d.
Yorkshire (except Cleveland)	..	...	...	48 to 56	......	$6\frac{1}{2}d.$ to 9d.
Lancashire and Cheshire	...	...	...	...	$48\frac{1}{2}$ to $55\frac{1}{2}$	......	$6\frac{1}{2}d.$ to $9\frac{1}{2}d.$

[1] At Maryport; the next lowest rate is $7\frac{1}{4}d.$

ENGLAND AND WALES (*Continued*).

	Working hours per week.	Rate per hour.
North and West Midland Counties	49½ to 59	6*d.* to 9½*d.*
South Midland and Eastern Counties	53 to 61½	5¾*d.* to 9*d.*
London District	50	10½*d.*
South-eastern Counties	54 to 68	6*d.* to 9½*d.*
South-western Counties	53 to 64	5½*d.* to 8*d.*
Wales and Monmouth	53 to 56½	5¾*d.* to 8½*d.*
SCOTLAND	51 to 57	6½*d.* to 10*d.*
IRELAND	54 to 60	5½*d.* to 8½*d.*
ISLE OF MAN	55	5*d.* to 6½*d.*

The following rules were made as from May 1st, 1900, between the London Master Builders' Association and the Carpenter and Joiners' Trade Society, viz.:—

1. That the working hours in summer shall be 50 per week for 40 weeks.

That during twelve weeks of winter, commencing on the second Monday in November, the working hours shall be for the first three weeks and the last three weeks 47 hours per week, and during the six middle weeks 44 hours per week. Joiners in shops to have one hour for dinner throughout the year.

2. That the present rate of wages shall be advanced one halfpenny per hour from May 1st, 1900 (making 10½*d.*).

3. That overtime, when worked at the request of the employers, shall be paid at the following rates, namely:—From leaving-off time until 8 p.m., time and a quarter; from 8 p.m. to 10 p.m., time and a half; after 10 p.m., double time. No overtime shall be reckoned until each full day has been made, except where time is lost by stress of weather. On Saturday the pay for overtime from noon to 4 p.m. shall be time and a half, after 4 p.m. and Sunday double time. Christmas Day shall be paid for the same as Sunday.

4. That one hour's notice be given, or one hour's time be paid, by either side on determining an engagement. All wages due shall be paid at the expiration of such notice, or walking-time if sent to yard. All workmen who are in receipt of full wages, and who have been employed for not less than 42 hours, shall, on discharge, receive one hour's notice, to be occupied, so far as is practicable, in grinding tools, with one hour's pay in addition. In the event of more than *ten per cent* of the workmen of the trade employed at the job giving notice to leave during any one day (except Saturday), they shall not be entitled to receive their money until noon on the following day.

5. That men who are sent from the shop or job, including those engaged in London and sent to the country, shall be allowed as expenses 6*d.* per day for any distance over six miles from the shop or job, exclusive of travelling expenses, time occupied in travelling, and lodging-money.

6. That payment of wages shall commence at noon, or as soon thereafter as practicable, on Saturday, and be paid on the job. But if otherwise arranged, walking-time at the rate of three miles per hour to get to the pay-table at twelve noon.

7. That employers shall provide, where practicable and reasonable, the following conveniences:—

(*a*) A suitable place for the workmen to have their meals on the works, with a labourer to assist in preparing them. (*b*) A lock-up where tools can be left at the owners' risk. (*c*) A grindstone for the use of workmen.

8. That wages earned after leaving-off time on Friday and Saturday only shall be kept in hand as back time.

9. That the term "London district" shall mean 12 miles radius from Charing Cross.

10. [This clause deals with the settlement of trade disputes by means of a Board of Conciliation composed of three members nominated by the employers and three nominated by the workmen, or if the dispute involves "claims or rights of other sections of the building trades", by a Joint Conciliation Board.]

11. That six months' notice on either side shall terminate the foregoing rules. (It is understood that the six months' notice shall not expire during the winter weeks.)

The foregoing rules give a good idea as to the arrangements between employer and employee in the London district, and also, it is believed, in the principal towns of the kingdom. Local customs, however, cause departures in some cases, viz., as to the amount of back time in hand, and also the day for paying wages. In some towns wages are paid on Friday, the back time consisting of the wages earned after leaving-off time on Thursday, and the rates as regards overtime allowance are non-existent in many places.

The hours for mill sawyers and wood-working machinists in the London district are

similar to those of the joiners, but in regard to overtime there is a slight difference. This, when worked at the request of employers, *but not otherwise*, shall be paid at the following rates, namely:—From leaving-off time until 6 p.m., *ordinary rate*; from 6 p.m. to 10 p.m., *time and a quarter*; after 10 p.m., *double time*. In other respects the allowance for overtime is similar to that already mentioned. In regard to the termination of an engagement, one hour's notice shall be given, or one hour's time be paid by either side.

CHAPTER III

TOOLS AND MACHINERY, INCLUDING HINTS ON USE, REPAIR, &c.

The principal tools used by the wood-worker are planes, chisels, and saws, to which may be added hammers, mallets, squares, screw-drivers, gimlets, brad-awls, pincers, oil-stone, slip and oil can, dividers and compasses, brace and bits, auger bits, spoke-shaves, levels, gauges, blocks, bench-hook, plumb and chalk lines, bench brush, &c.

The following table describes the planes in general use:—

Names	Length of Stock in inches.	Width of Stock in inches.	Breadth of Iron in inches.
Modelling-planes	1 to 5	$\frac{1}{4}$ to 2	to $1\frac{1}{2}$
Smoothing-planes	$6\frac{1}{2}$ to 8	$\frac{1}{2}$ to 3	$1\frac{3}{4}$ to 2
Rabbet or rebate planes (right and left)	$9\frac{1}{2}$	to 2	to 2
Jack-planes	12 to 17	$2\frac{1}{2}$ to 3	2 to $2\frac{1}{4}$
Panel-planes	$14\frac{1}{2}$	$3\frac{1}{2}$	$2\frac{1}{2}$
Trying-planes	20 to 22	$2\frac{1}{4}$ to 3	2 to $2\frac{1}{2}$
Long-planes	24 to 26	3	2
Jointer-planes	28 to 30	$3\frac{3}{4}$	$2\frac{3}{4}$
Cooper's jointer-planes	60 to 72	5 to $5\frac{1}{4}$	$3\frac{1}{2}$ to $3\frac{3}{4}$

Such planes as hollows and rounds, matching and moulding planes are usually shop planes furnished by the employer and used in common, although one occasionally comes across a workman whose kit also comprises a set of these.

The general angles and purposes of ordinary plane-irons are nearly as follows:— Common pitch or 45° from the horizontal line is used for all bench-planes for deal and similar soft woods; York pitch or 50° for bench-planes for mahogany, wainscot, and hard or stringy woods; middle pitch or 55° for moulding-planes for deal, and smoothing-planes for mahogany and similar woods; half pitch or 60° for moulding planes for mahogany and woods difficult to work. When the iron is perpendicular, or leans slightly forward, the tool is called a scraping-plane, and with it close hard woods may be smoothly scraped if not cut. A toothing-plane has a perpendicular iron, which is grooved on its face to present a series of fine teeth instead of a continuous edge, and it is employed for roughing veneers and the surfaces to which they are to be attached.

An outfit of chisels will comprise a set of firmer-chisels from $\frac{1}{8}$ to 2 inches, two long firmer-chisels for paring ($1\frac{1}{4}$ and $1\frac{3}{4}$ inch), two socket-chisels for heavy work ($\frac{3}{4}$ and $1\frac{1}{2}$ inch), bench-gauges from $\frac{1}{2}$ to 2 inches, and an inch blunt scraping-chisel. All should be kept in order in racks at the back of the bench and handled ready for immediate use.

Of saws, there will be required one rip and one cross-cut hand-saw, panel-saw, back-saws, compass-saw, pad-saw, frame-saw, and others of special character.

The following table describes the sizes and gauges of saws in general use:—

Names			Length in inches.	Breadth in inches.		Teeth per inch.	Thickness.	
				Handle.	Point.		Inch.	Gauge.
Taper, without Backs—								
Rip-saw	...	...	28 to 30	7 to 9	3 to 4	3 to 3½	·04 to ·05	18 or 19
Half rip-saw	...	...	26 to 28	6 to 8	3 to 3½	4	·04 to ·05	18 or 19
Hand-saw	...	...	10 to 26	5 to 7½	2½ to 3	5	·04 to ·05	18 or 19
Panel-saw	...	...	10 to 26	4 to 7½	2 to 2½	7	·035 to ·04	19 or 20
Chest-saw	...	...	10 to 20	2½ to 3½	1¼ to 2	6 to 8	·032 to ·04	18 to 21
Table-saw	...	...	18 to 26	1¾ to 2¼	1 to 1½	7 to 8	·04 to ·065	16 to 19
Compass-saw	...	...	8 to 20	1 to 1½	½ to ¾	8 to 9	·04 to ·05	18 or 19
Keyhole-saw	...	...	6 to 12	½ to ¾	to ¼	9 to 10	·035 to ·04	19 or 20
Parallel, with Backs—								
Tenon-saw	...	...	12 to 20	3 to 4		10	·028 to ·032	21 or 22
Dovetail-saw	...	...	6 to 10	1½ to 2		14 to 18	·02 to ·022	24 or 25

The dulled edges of cutting tools, such as chisels and planes, are restored to keenness by successive actions of the grindstone and oil-stone, the former being employed to remove the bulk of the material in order to prepare the tool for the slower but more delicate action of the oil-stone. The grindstone should be kept in order, as far as possible, by an equal distribution of wearing. Narrow tools should be traversed across the face of the stone to avoid wearing the latter into ridges, and the edges of the stone should also have their fair amount of work, otherwise the stone will become hollow and unfit for grinding broad flat tools. The operation of setting a cutting tool is the formation of two flat facets inclined to one another at a certain angle, their intersection forming the cutting edge. In grinding wood-working tools it is well to stop before a feather edge is produced, as this has to be removed, with a consequent loss in the length of the iron. Tools should be ground with their edges parallel to the axis of the grindstone, so as to make the ground bevel concave to the same radius as the stone, and therefore better prepared for the action of the oil-stone.

Glue is an indispensable adjunct in forming joints in the finer kinds of work. It may be made waterproof by the addition of bichromate of potash, which should be previously dissolved in water. The hotter glue is when applied, the greater will be its power in holding the surfaces together, therefore glue in all large joints should be applied as soon as possible after boiling. Glue will not set in a freezing temperature, the setting being prevented by great cold. In melting ordinary glue in the common form of glue-pot, it is a good plan to add salt to the water in the outer vessel. It will not boil then until heated considerably above its ordinary boiling-point, and the heat is retained longer.

Modern mills are now often fitted with band-saws in addition to circular saws, the principal advantages claimed for the former being that, whilst cutting as fast as circular saws, they waste only about a quarter as much wood in each cut—an important item in hard-wood conversion—and take considerably less power to drive; they also allow the sawyer to see how the logs are developing, thus enabling him to convert to the greatest advantage.

In the selection of saws great care has to be exercised, the saw to be purchased being the one that does the best work in the greatest quantity with a minimum amount of trouble. Sawyers have generally individual opinions as to the tempering, sharpening, setting, packing, and working of saws, and also as to the gauge and form of teeth. Experience alone will teach an intelligent man how to keep his saw blades in order and work them to the best advantage, and here a competent man saves his wages many times over in the saving effected of time and material. For a saw tooth to give good results it must not only do its work well, but must be easily sharpened, and consequently of such a shape that the file or emery

stone (as the case may be) can readily get at it. Whatever the form of teeth may be, the gaps or gullets between them should never be less in area than the teeth themselves in order that there may be ample space for the saw-dust to lie in, until an opportunity occurs of disengaging itself. The tempering of a saw is also an important matter. The essential points are that the temper should be of an even nature throughout the surface of the saw, that it should be properly hammered and not too hard for a single-cut file to be readily used on it. For cutting soft wood the teeth should be larger, spaced farther apart, and be more raked than when working on hard wood. The teeth of frame and all reciprocating saws should be less raked than those of circular blades when doing the same class of work, and for band-saws, although the teeth are generally smaller, they should be (when a gulletted tooth is used) of much the same shape as those of the frame saw. The proper gauge for a saw of any description is the thinnest at which it will do its work properly, and there are certain usual gauges according to the size and class of saw and the work required of it, which can be ascertained on reference to the catalogues.

As the satisfactory results from all wood-working machines depend so greatly on the saws or cutters being kept in good condition, it is evident that too much care and attention cannot be bestowed on this point.

The sharpening of machine cutters is usually done by means of an ordinary grindstone revolving in a trough; the larger the grindstone is in diameter the better, and it should not be less than 8 inches on the face. It should be kept as round as possible, and for this reason no tools but those used for the machines should be ground on it. It is also usual to fit the end of the spindle with a revolving set-stone for finishing off the cutters as they leave the larger stone. In large establishments automatic grinding machines are commonly used.

The bevel to which the cutting edge of a plane-iron should be ground depends on the class of work it has to do; the softer the wood, the longer the bevel. When working soft woods the angle formed by the line of bevel and the face of the cutter should not be less than 25°; when working hard woods the angle should not be more than 40°.

Moulding-irons ought not to be too wide. It is better to work a wide moulding with a number of narrow irons with few curves in each than to do it with one or a few wide cutters ground in a complicated manner. Not only are the narrow irons more easily sharpened in the first case, but, should by any chance the cutting edge get damaged, it is far simpler and cheaper to remove and sharpen a narrow iron than a wide one.

Saws and moulding and plane irons are generally kept in cupboards in the grinding rooms, and where chain-mortising machines are used, the chains are usually kept in oil in shallow trays.

Where steam is the motive power in use, incrustation and scale in the boilers are often of serious consequence, as nearly all waters contain foreign substances. The nature and hardness of the scale formed will depend upon the kind of substances held in solution and suspension. Analyses show that carbonate and sulphate of lime form the larger part of all ordinary scale, that from carbonate being soft and granular, and that from sulphate hard and crystalline. Organic substances in connection with carbonate of lime also make a hard and troublesome scale. The presence of scale or sediment in a boiler results in loss of fuel, burning and cracking of the boiler, predisposes to explosion, and leads to extensive repairs. It is absolutely essential to the successful use of any boiler, except with pure water, that it be accessible for the removal of scale, for though a rapid circulation of water will delay the deposit, and certain chemicals will change its character, yet the most certain cure is periodical inspection and mechanical cleaning. This may, however, be rendered less frequently necessary, and the use of very bad water more practical, by the employment of some preventive, which ought to be selected under expert advice after the water has been analysed. When boilers are coated with a hard scale difficult to remove, it will be found that the addition of $\frac{1}{4}$ lb. of caustic soda per horse-power, and steaming for some hours, according to

the thickness of the scale, just before cleaning, will greatly facilitate that operation, rendering the scale soft and loose. This should be done, if possible, when the boilers are not otherwise in use.

For the lubrication of machinery mineral oils should only be used with discrimination and under exceptional circumstances. Neat's-foot, lard, and olive-oil are to be recommended. "Stauffers" lubricant may be used to advantage for solid grease lubricators, which are more economical than the needle type, and are supplanting the latter for shafting.

For all ordinary purposes leather belting is the best for saw-mill requirements, but it should be of the very best quality. For repairing belts Green's fasteners are the kind generally used, as they make perfectly flat joints, and consequently the belts run very smoothly; they are easily attached and uncoupled.

For much of the information given, the writer has referred to *The Woodworker's Handybook*, by P. N. Hasluck, and *Modern Woodworking Machinery*, by J. S. Ransome.

CHAPTER IV

PRIME COST, BOOK-KEEPING, &c.

To arrive at the prime cost of any work, it will be necessary to collect the cost for labour and materials, and add a percentage for wear and tear of machinery, office and other trade expenses, rental, and personal profit. It is advisable for each man to fill in and return a time sheet, for which pattern A will serve as a model. The division of labour on this

A.—TIME SHEET

TIME SHEET.—Week Ending 6th November C. CHIPS—Carpenter			
J. SAWDUST & SON, Builders.			*Materials to be entered on back.*
Where Employed.		Hours.	Description of Work.
Saturday.	Shop	5	Making sashes and frame. B/1 (Mr. Brown's).
Monday.	Shop	1	Do. B/1
Tuesday.			
Wednesday.			
Thursday.			
Friday.			
	Total Hours......	6	@ 10½d. per hour. Wages paid, £0, 5s. 3d.

sheet should be abstracted into a "Cost Labour" book, ruled as B, each particular job on the time sheet being allocated to its proper column in this book. Where an extensive business is carried on the columns in this book will require to be very many, and for practical purposes a book with, say, one hundred lines ruled to a page, and twenty cash columns ruled across it, might be found sufficient. If only twenty men are employed such a page would

last, say, four weeks, whilst if the number exceed one hundred it would be necessary to continue the abstract on to another page.

The material can be taken from the foreman's setting-out sheet or board, or from the

B—COST LABOUR BOOK

Week Ending	Name of Workman.	Trade.	Mr. Brown.																Total.		
			£	s.	d.	£	s.	d.	£	s.	d.	£	s.	d.	£	s.	d.	£	s.	d.
6th Nov. D/1	Chips....... Sawyer.....	Carpenter..... Machinist.....	0 0	5 1	3 6	... ...	... ...	... ...	... ...	... ...	... ...	... ...	... ...	... ...	... ...	... ...	... ...	0 0	5 1	3 6

time sheet when the joiner has made a note of it there, and entered into a "Cost Material" book, ruled as C, in which it should be valued at an average cost rate delivered into the yard. The totals should be entered from the "Labour" and "Material" books week by

C—COST MATERIAL BOOK

Week Ending	Description of Materials.	Rate.				Total.		
		d.	£	s.	d.	£	s.	d.
6th Nov.	MR. BROWN'S WINDOW							
	12' 0" sup. 1" deal.....................	$2\frac{1}{2}$	0	2	6			
	7' 0" „ 2" „	$4\frac{1}{2}$	0	2	7			
	6' 0" „ $1\frac{1}{4}$" „	3	0	1	6			
	8' 0" „ $\frac{1}{2}$" „	1	0	0	8			
	15' 0" run $\frac{3}{8}$" bead....................	$\frac{1}{4}$	0	0	4			
	15' 0" „ $1\frac{1}{4}$" × $\frac{3}{4}$" bead...........	$\frac{1}{4}$	0	0	4			
	3' 9" „ 6" × 3" oak sill..........	8	0	2	6			
D/1	Glue, blocks, and wedges............	...	0	0	6			
	2 hours, machines....................	2/9	0	5	6			
						0	16	5

week into a "Contract" book, ruled as D. The addition of the amounts for labour and materials, plus a percentage for working and other expenses, will, if the total be compared with the amount of contract, show the result of either "profit" or "loss".

D—CONTRACT BOOK

Week Ending		Labour.			Materials.			Amount of Contract.		
		£	s.	d.	£	s.	d.	£	s.	d.
6th Nov.	To amount of Contract for making Sashes and Frame for Mr. Brown..							1	5	6
	By net amount of Labour.....................................B/1	0	6	9						
	„ „ „ Materials.......................C/1				0	16	5			
		0	6	9	0	16	5			
	Labour............				0	6	9			
	Profit.............				0	2	4			
					1	5	6	1	5	6

The amount of contract when completed would be carried to the Ledger, and the account rendered in the usual way. It will be found of great advantage to reference all entries,

thus time entered from the time sheet should be referenced on the sheet to the folio of the "Labour" book in which it may be found; material should be similarly treated with the "Material" book; entries from these books into the Contract book, and from the last into the Ledger should all be referenced. A considerable saving of time will be effected, when inquiries have to be made, if this is done.

The system above advocated gives the net total cost of each job. If, however, the actual prime cost of any particular portion of the work is required, it will be necessary to take especial notes of the labour spent and material used.

Thus, supposing the prime cost is required for a deal cased frame (3 × 6 feet sight size), having 1-inch inside and outside linings, 2-inch heads, $1\frac{1}{4}$-inch pulley stiles, $\frac{1}{2}$-inch back linings and parting slips, $\frac{3}{8}$-inch parting beads, $1\frac{1}{4} \times \frac{3}{4}$-inch inside bead, 2-inch moulded double-hung sashes, and 6 × 3-inch oak sill, the stuff required would be as follows:—

	£	s	d
3-feet 9-inch run, 6 × 3 inches, English oak sill at 8d.	£0	2	6
12-feet superficial 1-inch deal (linings) at $2\frac{1}{2}d$.	0	2	6
2-feet superficial 2-inch deal (head) at $4\frac{1}{2}d$.	0	0	9
5-feet superficial 2-inch deal (sashes) at $4\frac{1}{2}d$.	0	1	10
6-feet superficial $1\frac{1}{4}$-inch deal (pulley stiles) at 3d.	0	1	6
8-feet superficial $\frac{1}{2}$-inch deal (back linings and slips) at 1d.	0	0	8
15-feet run, $\frac{3}{8}$-inch parting-bead, at $\frac{1}{4}d$.	0	0	4
15-feet run, $1\frac{1}{4} \times \frac{3}{4}$-inch inside bead, at $\frac{1}{4}d$.	0	0	4
Glue, blocks, and wedges, say	0	0	6
Machine and machinist (cutting out, planing, and moulding), say 2 hours at 3s. 6d.	0	7	0
Joiner's time, putting together and fitting sashes, say 6 hours at $10\frac{1}{2}d$.	0	5	3
	£1	3	2
Add 10 per cent	0	2	4
	£1	5	6

This divided by area of frame (23 feet) will give an average of 1s. $1\frac{1}{4}d$. as the cost per foot super for above sashes and frames. The writer is well aware that sashes and frames can be obtained for 6d. per foot super and less, but in such cases the deals have been bought for considerably less than £20 per standard. The principal reasons that account for this cheap joinery are as follows:—*Firstly*, the purchase and use of timber in short lengths and small scantlings, which can be bought at very low rates, in fact at almost firewood prices; *secondly*, the labour employed in putting together is in great part that of improvers; *thirdly*, the adoption of piece-work at cutting prices.

Where a large quantity of similar articles have to be made the work would be divided, and in such a case averages of the result of labour should be taken.

Where the material is entered on a board by the "setter out" previous to giving it to the machinists to prepare the stuff, it will be found a very good plan to give the board a coat of quick-drying varnish; this preserves the figures, and, further, should any mistake arise, there will be no difficulty in fixing on the responsible party. After use these boards should be kept for a reasonable time in the mill office.

Planks, deals, battens, boards, &c., are carried by rail at computed and not actual machine weights, as follows:—Rough and unplaned planks, &c. (excepting pitch pine and hard woods), are always reckoned on a basis of $2\frac{1}{2}$ tons per standard; pitch pine deals, &c., under 4 inches thick, on a basis of 3 tons per standard; planed boards (but not planks) any thickness (excepting pitch pine and hard woods), on a basis of 2 tons 2 cwts. per standard; birch, oak, ash, elm, mahogany, teak, beech, greenheart, hickory, and round timber generally are computed as weighing 40 cubic feet to the ton; pitch pine, spruce, whitewood, elm, redwood, walnut, maple, pine, and fir timber are computed as weighing 50 cubic feet to the ton.

Section XIV.—ESTIMATING

BY

W. E. DAVIS

AUTHOR OF "QUANTITIES AND QUANTITY TAKING"

CHAPTER I

QUANTITIES: TAKING-OFF

In these days when the spirit of commercialism enters so largely into every business transaction, comparatively few people give orders for work regardless of the final cost, and it is therefore necessary to find some system of arranging for a definite cost of anticipated work, so that this definite sum will form a basis of agreement between the employer and the contractor. As a general rule the contractor arrives at this sum by pricing a "Bill of Quantities", containing full detailed measurements of all materials and labour in the proposed works, as shown on the drawings and described in the specifications. The bill must be in such a form as to cover practically everything included in the work, and in such detail that comparatively little is left to the imagination of the person estimating.

In the preparation of a Bill of Quantities the first requirements are carefully-prepared drawings and a carefully-written specification. These, of course, belong to the province of the architect and not to the surveyor, who in some cases, it must be confessed, has to do the best he can with very slight information as to the requirements and the intentions of the architect. When such is the case, however conscientious the surveyor may be, he has no alternative but to make himself and the contractor "safe" by leaning always to the more expensive work, thus covering the cost of what the architect, during the execution of the work, may order with no intention of departing from the contract. This, it will readily be seen, is prejudicial to the interests of the employer, as in many cases, in order to make his quantities inclusive, the surveyor will probably overleap the mark.

Where the quantities form part of the contract, as they frequently do now, this objection loses *some* of its weight but not all, as the fact of the quantities forming part of the contract leaves the architect free to have the work as executed remeasured, but in this case the employer has to pay for the remeasurement as well as for the original quantities.

In the preparation of Bills of Quantities there are three principal stages:—*Firstly*, "Taking off" the dimensions from the drawings; *secondly*, "Abstracting" the dimensions under their various headings and trades, for which purpose they are previously worked out or squared; and *thirdly*, "Billing", or making the abstracted dimensions (after being cast together) into a bill, which is the finished work as far as the surveyor is concerned; the subsequent pricing is usually performed by the contractors tendering.

In each of these stages, but principally in the first and third, there are a good many points to be observed to arrive at the best result, not as to price merely, but as to the arrangement to facilitate any subsequent work in the variations that may arise upon the contract. The abstracts should as nearly as possible anticipate the subsequent form of the

bill in order and arrangement, although frequently at the "Billing" stage it will be found advantageous to make some slight variation in form and arrangement which would not be apparent at the "Abstracting" stage. The work is gone through by a second person and checked at the various stages, except the "Taking off". The correctness of this lies entirely with the person taking off.

As it frequently happens that the dimensions will be referred to at later stages, *i.e.* during the progress of the contractor's work and at its completion, it behoves the surveyor to make them thoroughly explanatory, so that if the surveyor who took off the dimensions is not at liberty to take up the variations, they shall be perfectly clear to anyone endeavouring to trace the work taken and included in the quantities. For this purpose numerous headings will be necessary covering the main items; sub-headings giving the subdivisions of these main items, and notes to the dimensions themselves locating them exactly. Only those surveyors who have had to deal with dimensions with and without these headings and notes can appreciate the value in saving of time and trouble by their use.

As to the order in taking off, it matters little which is adopted, provided that every item is perfectly clear, but it is well for each surveyor to follow a definite arrangement and thus avoid the possibility of leaving out some items entirely.

The system of measurement varies in different parts of the country, but the London system, having by far the larger number of devotees, will be that selected for the examples given in this work. The variations between the different systems are comparatively slight, and that here given will be a good general basis for the commencement of operations. Moreover, contractors have of late years become so used to London quantities, that they have little trouble in arriving at a satisfactory estimate when this system has been adopted.

The foregoing remarks as to the points to be borne in mind in the preparation of quantities, so that they may be of most value in arriving at a close estimate, free from any undue element of speculation, and so that the settlement of accounts at the completion of the work will be as easy as possible, may be summarized as follows:—

1. The dimensions to be taken in such a form that they may be easily referred to in the event of future enquiries.

2. That, providing the drawings and specifications supplied are sufficiently explanatory— as they should be—the items taken shall exactly cover the work intended, without favour to either the employer or the contractor.

3. The bill to be written in such a form as to make it possible to locate as nearly as possible the various items, and the descriptions to tally exactly with those in the specification.

The first stage in the preparation of a Bill of Quantities is "taking-off" the dimensions from the drawings, and while a few general instructions may be useful, a great deal more information will be gained by careful examination of the examples given, tracing the dimensions one by one upon the drawings. These examples, while by no means exhaustive, will give a good basis for the study of the subject.

As an example of carpenters' work, a somewhat elaborate roof (fig. 1196) has been selected as affording a rather extensive variety of items—the eaves being further varied to give a still greater variety. With this example the system of measuring will be quite obvious. As a general rule, the main constructional portion is billed at *per foot cube*, but any items less than 4 inches in width should be billed at *per foot run*, and anything of or over this width, but less than 2 inches in thickness, at *per foot superficial*. The allowances for "scarfings" are taken where timbers are more than 25 feet long. A rough rule is to allow three times the depth of the timber as an extra length. The surveyor should be continuously on the look-out for any small items that may have been unintentionally omitted from the drawings. An instance of this is given in the case of the bridging between rafters for fixing the upper and lower edges of the ceiling boarding. Instructions as to spacing of rafters, &c., are generally given in the specification. In the

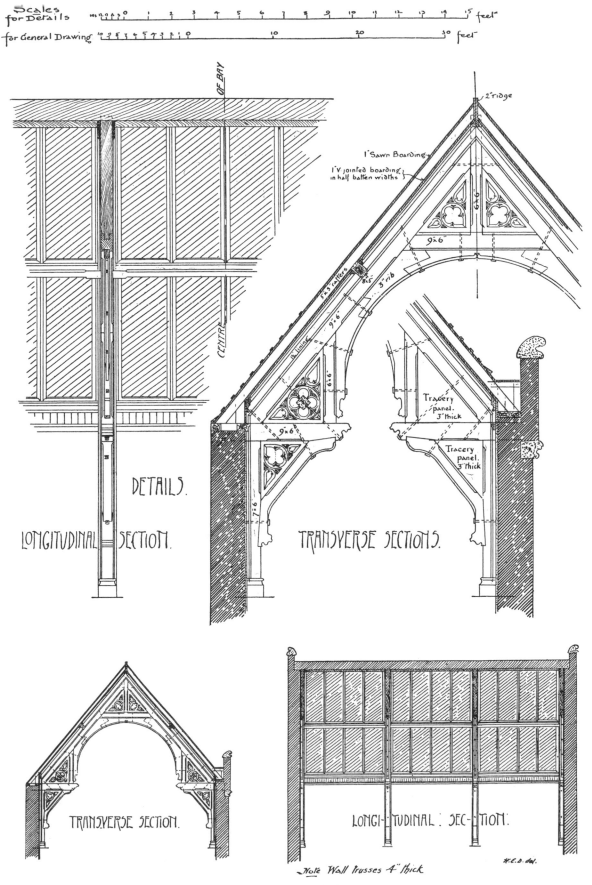

Fig. 1196.—Roof of Church

example they are taken at the usual spacing, *i.e.* 12 inches apart, allowing for one at each side of the principal for fixing the ceiling boarding to. The various sizes and thicknesses are generally mentioned in the specification, but in the example given are figured on the drawing. It will be obvious that, as the ceiling boarding is laid diagonally, there will be at every edge a certain amount of waste; an allowance of 3 inches at each edge is made for this. A similar allowance would be made for any raking cutting to ordinary boarding, *i.e.* at hips and valleys.

The fall of the gutter should be mentioned in the specification, and the roof-plan would give the number of drips and the position of the cesspool, and *should* show the proper width. The surveyor should, however, work this out for himself. In the example given 9 inches is taken as the minimum width, the drips 8 feet apart and the cesspool at the end. The gutter is calculated to fall 2 inches in 10 feet and the drips are 2 inches deep. This gives the average width of 1 foot 3 inches as taken.

The lintels and centring are taken with the joiner's work, as these would be measured with the openings. The usual thickness for lintels is one inch in depth for every foot of opening, with a minimum depth of 3 inches. The arches being flat and 4½ inches on soffit, a turning-piece the *width* of the opening is taken. If the arches had such a rise that a solid turning-piece would be impracticable, the girth of the arch would be taken, and anything 9 inches in width and over would be billed at *per foot superficial*, and anything under this width at *per foot run*, stating the form of the arch, *i.e.* pointed, semicircular, elliptical, segmental, or four-centre-pointed.

In measuring joiner's work, a general rule is to measure everything over 4 inches in width at *per foot superficial*, but if there is much labour to the items, it is better to keep them at *per foot run*, and this is the system adopted here. Panelling should be measured at *per foot superficial*, noting the disposition of the panels, and any extra work to angles, cappings, plinths, &c., must be taken separately. Any information as to girth of mouldings, and any other particulars that will enable the contractor to more clearly estimate the cost of the work, should be given wherever possible. If the thicknesses are not to be net as specified, a note must be made to this effect, otherwise *nominal, i.e.* (after an allowance has been made for sawing and planing) will be assumed.

The examples (figs. 1197 and 1198) call for no comment, as every dimension can be traced upon the drawings. For explanation of dimensions see note, page 394.

			ROOF (fig. 1196)					7″
								4″
			2 Wall Trusses.					—
2/2/	7·0			2/2/	7·0	25·8		·11
	·7				·11			=
	·4			2/2/	5·9	32·7		9″
		5·5	Pitch-pine (*or other material as Speci-*		1·5		Wrought face on pitch-pine.	2/4″ = 8″
			fication) framed in roof trusses	2/2/	4·6	21·0		1·5
2/2/	5·9		*wall-posts.*		1·2			=
	·9					79·3		6″
	·4	5·9	Add *hammer beam.*					2/4″ = 8″
2/2/	4·6							1·2
	·6							=
	·4	3·0	Add *posts.*					9″
2/2/	16·6							4″
	·9			2/2/	16·6	71·6		1·1
	·4	16·6	Add *principals.*		1·1			=
2/	7·0			2/	7·0	19·10	Add	9″
	·9				1·5			2/4″ = 8″
	·4	3·6	Add *tie.*	2/	5·3	12·3		1·5
2/	5·3				1·2			=
	·6					103·7		6″
	·4	1·9	Add *post.*					2/4″ = 8″
		35·11						1·2
								=

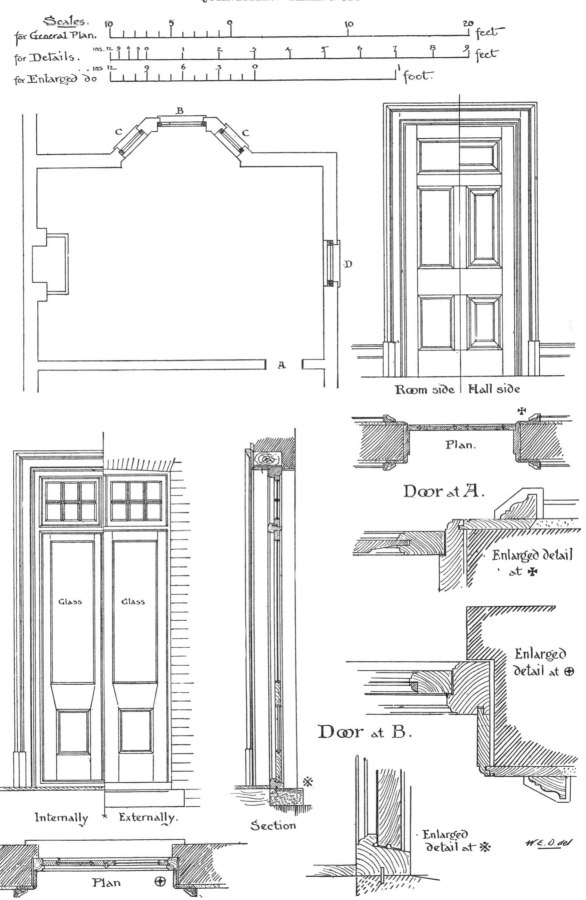

Fig. 1197.—Plan of Room and Details of Joinery

2/2/	2·0		
		8·0	Stopped chamfer 1½″ wide, on pitch-pine *wall-posts.*
2/2/	1·6		
		6·0	Add *hammers.*
2/2/	2·9		
		11·0	} Add *posts.*
2/2/	3·3		
		13·0	
2/2/	3·6		
		14·0	} Add *principals.*
2/2/	3·9		
		15·0	
2/2/	2·9		
		11·0	
2/	5·9		
		11·6	Add *tie.*
2/2/	2·3		
		9·0	Add *post.*
		98·6	
2/17/2/	= 68		Rounded stops to 1½″ chamfer.
2/2/	2·6		
		10·0	Stopped chamfer 3″ wide, on pitch-pine *tie.*
2/2/	= 4		Square stops to ditto.
2/2/	= 4		Moulded ends to 9″ × 4″ pitch-pine hammer beams, as sketch.
2/2/	= 4		Pitch-pine wrought on one face and edges, and framed brackets, as sketch, out of 18″ × 4″ and 6 ft. long, stop-chamfered on two edges *(under hammer beam).*

2/2/	5·4 / 1·3		
		26·8	} 2½″ pitch-pine (*or other material*), wrought one side, and framed ribs, in extra wide timber, including scarfings (cutting measured).
2/2/	4·0 / 1·2		
		18·8	
2/	6·3 / 1·2		
		14·7	
		59·11	
2/2/	2·3		
		9·0	} Raking, cutting on ditto.
2/2/	3·9		
		15·0	
2/2/	2·10		
		11·4	
2/2/	1·3		
		5·0	
		40·4	
2/	18·6		
		37·0	Circular wrought, cut, and chamfered (1½″ wide) edge to 2½″ rib.
2/1/	= 2		Extra to break in ditto, as sketch.
2/2/	= 4		Moulded ends to 2½″ rib, including long stop to 1½″ chamfer, as sketch.
2/2/	3·9		
		15·0	} Stopped rebate, 3½″ girth, in pitch-pine *for rib.*
2/2/	4·9		
		19·0	
2/	6·8		
		13·4	
		47·4	

2/2/	= 4		2½″ pitch-pine triangular perforated tracery panels, 2′ 3″ × 2′ 8″ extreme, cusped and worked one side, as sketch, cross-tongued as required, "A".
2/2/	= 4		Ditto, but 2′ 6″ × 3 ft. extreme.
2/2/	= 4		Ditto, but 2′ 11″ × 3′ 6″ extreme, and as sketch, with carved boss in centre, "B".
2/2/	2·3		
		9·0	
2/2/	2·8		
		10·8	
2/2/	3·6		
		14·0	
2/2/	2·6		
		10·0	
2/2/	3·0		
		12·0	} Stopped rebate, 3½″ girth, in pitch-pine *for tracery panels.*
2/2/	3·10		
		15·4	
2/2/	2·11		
		11·8	
2/2/	3·6		
		14·0	
2/2/	4·6		
		18·0	
		114·8	
		9″	
		6″	
		1·3	
2/2/	1·3		
		5·0	10″ × 1½″ pitch-pine plinth to posts, planted on in short lengths, and moulded 5″ girt.
2/2/1/	= 4		External mitres.
2/2/2/	= 8		Fitted ends.
2/2/	7·3		
		29·0	2″ × 1½″ pitch-pine chamfered, angle fillet planted on *on wall-posts.*
2/2/1/	= 4		Ends scribed over moulded plinth.
2/1/	= 2		Scribings over moulded wall-plate, 8″ girt.
2/2/1/	= 4		Mitred ends.
2/2/	7·9		
		31·0	} 7″ × 1½″ pitch-pine chamfered, lining planted on *on side of principal.*
2/2/	8·0		
		32·0	
2/2/2/	= 8		Fitted ends.
2/2/1/	= 4		Ditto on splay.
2/1/	= 2		Irregular mitre.

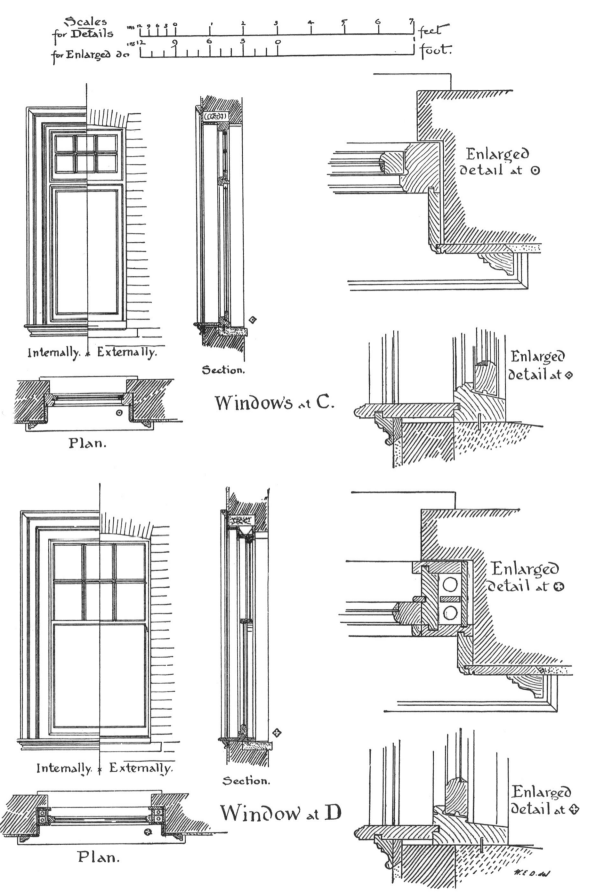

Scales
for Details
for Enlarged do

feet
foot.

Internally. & Externally.

Section.

Enlarged detail at ⊙

Enlarged detail at ⊕

Windows at C.

Plan.

Internally. & Externally.

Section.

Enlarged detail at ✪

Window at D

Enlarged detail at ⊕

Plan.

Fig. 1198.—Details of Joinery (see fig. 1197)

2/1/		= 2	Hoisting and fixing roof truss, 4" thick, 20 ft. span, and 20 ft. rise, the ridge ...ft. above ground.	2/34/2/	2·6	= 136	Rounded stops to 1½" chamfer.
2·2/2/2/		= 16	Fixing ¾" bolt 12" long, and boring pitch-pine.	2/2/2/		20·0	Stopped chamfer 3" wide, on pitch-pine.
2·2/2/		= 8	Ditto, 1' 8" long, and ditto.	2/2/2/		= 8	Square stops to ditto.
2·2/2/2/		= 16	Ditto, 2 ft. long, and ditto.	2/2/		= 4	Moulded ends to hammer beams, as before, but 9" × 6".
2·2/2/		= 8	Ditto, 4 ft. long, and ditto.	2/2/		= 4	Pitch-pine brackets, as previous sketch, but wrought on both face and edges out of 18" and 6", and stop-chamfered on four edges.
2·2/2/		= 8	Ditto, 1' 6" long, and ditto.				
2·2/2/		= 8	Ditto, 2' 2" long, and ditto.	2/2/	5·4 / 1·3	26·8	3" pitch-pine (*or other material*) wrot both sides, and framed ribs, in extra wide timbers, including scarfings (cutting measured).
2·2/2/		= 8	Ditto, 1' 2" long, and ditto.	2/2/	4·0 / 1·2	18·8	
2·2/1/		= 4	Ditto, 2' 6" long, and ditto.	2/	6·3 / 1·2	14·7	
			2. *Added to foregoing bolts for intermediate trusses.*			59·11	
2/2/	7·0 / ·7 / ·6	8·2	2 INTERMEDIATE TRUSSES. Pitch-pine (*or other material*) framed in roof trusses *wall-posts.*	2/2/	2·3	9·0	Raking, cutting on ditto.
2/2/	5·9 / ·9 / ·6	8·8	Add *hammer beams.*	2/2/	3·9	15·0	
2/2/	4·6 / ·6 / ·6	4·6	Add *posts.*	2/2/	2·10	11.4	
2/2/	16·6 / ·9 / ·6	24·9	Add *principals.*	2/	1·3	5·0	
2/	7·0 / ·9 / ·6	5·3	Add *tie.*			40·4	
2/	5·3 / ·6 / ·6	2·8	Add *post.*	2/	18·6	37·0	Circular wrot, cut, and twice chamfered (each 1½" wide) edge to 3" rib.
		45·10		2/1/		= 2	Extra to break on ditto, as sketch.
2/2/	7·0 / 1·8	46·8	Wrot face on pitch-pine.	2/2/		= 4	Moulded ends to 3" rib, as previous sketch, including two long stops to 1½" chamfer.
2/2/	5·0 / 2·6	57·6		2/2/	3·9	15·0	Stopped groove, 5" girt *for rib.*
2/2/	4·6 / 2·0	36·0	2/7" = 1·2 / ·6 = 1·8	2/2/	4·9	19·0	
2/2/	16·6 / 2·0	132·0	2/9" = 1·6 / 2/6" = 1·0 = 2·6	2/	6·8	13·4	
2/	7·0 / 2·6	35·0	2/2/6" = 2·0			47·4	
2/	5·3 / 2·0	21·0	2/9" = 1·6 / ·6 = 2·0	2/2/		= 4	Pitch-pine triangular perforated tracery panels, 2' 3" × 2' 8", as "A", but 3" thick, and cusped and worked on both sides.
		328·2	2/9" = 1·6 / 2/6" = 1·0 = 2·6	2/2/		= 4	Ditto, 2' 6" × 3 ft., ditto, ditto.
2/2/2/	2·0	16·0	Stopped chamfer 1½" wide, on pitch-pine *wall-posts.*	2/2/		= 4	Ditto, 2' 11" × 3' 6", ditto, ditto, with carved boss, as "B".
2/2/2/	1·6	12·0	Add *hammers.*	2/2/	2·3	9·0	
2/2/2/	2·9	22·0	Add *posts.*	2/2/	2·8	10·8	
2/2/2/	3·3	26·0		2/2/	3·6	14·0	
2/2/2/	3·6	28·0		2/2/	2·6	10·0	
2/2/2/	3·9	30·0	Add *principals.*	2/2/	3·0	12·0	Stopped groove, 5" girth *for tracery panels.*
2/2/2/	2·9	22·0		2/2/	3·10	15·4	
2/2/	5·9	23·0	Add *tie.*	2/2/	2·11	11·8	
2/4/	2·3	18·0	Add *posts.*	2/2/	3·6	14·0	
		197·0		2/2/	4·6	18·0	
						114·8	
				2/2/3/	·9	9·0	10" × 1½" pitch-pine plinth to posts, as before.
				2/2/2/		= 8	External mitres and fitted ends.

2/2/2/	7·3	58·0	2″ × 1½″ pitch-pine chamfered angle fillet, as before *on wall-posts.*
	2/2/2/	= 8	Ends scribed, as before.
	2/2/	= 4	Scribings over moulded wall-plate, as before.
	2/2/2/	= 8	Mitred ends.
2/2/2/	7·9	62·0	7″ × 1½″ pitch-pine chamfered lining, planted on *on side of principal.*
2/2/2/	8·0	64·0	
		126·0	
	2/2/2/2/	= 16	Fitted ends.
	2/2/2/	= 8	Ditto on splay.
	2/2/	= 4	Irregular mitres.
	2/1/	= 2	Hoisting and fixing roof truss as before, but 6″ thick. NOTE.—*Bolts added.*
			RAFTERS AND COVERING
29/	18·0 ·5 ·3	54·5	Fir-framed roof *rafters.*
29/	17·0 ·5 ·3	51·4	
			32·0 Scarfing 2·0 34·0
	34·0 ·11 ·2	5·2	Add *ridge.*
		110·11	
	2/		Fixing ¾″ bolts 11″ long, and boring.
			32·0 Scarfing 1·0 33·0
	33·0 ·5 ·4	4·7	Fir-framed roofs *plates.*
	33·0 ·6 ·3	4·2	
		8·9	
	33·0 ·6 ·3	4·2	Fir plate *outer, next parapet.*
15/	1·6 ·4 ·4	2·6	Fir-framed roofs in short lengths *cross ties.* NOTE.—*One taken each side of truss, and three intermediate in each bay.*
29/	1·9 ·5 ·3	5·3	Add *uprights.*
29/	1·3 ·5 ·3	3·9	
		11·6	
			32·0 Scarfing 1·0 33·0
	33·0 ·7 4	6·5	Pitch-pine framed roof *inner plate.*

	32·0		Wrot and hollow moulded edge to ditto, 8″ girt. 32·0 Scarfing 2·0 34·0
2/	34·0 ·8 ·5	18·11	Pitch-pine (*or other material*) framed roofs *purlins.*
2/	34·0 1·9	119·0	Wrot face, on pitch-pine.
2/3/2/	8·6	102·0	Stopped chamfer 1½″ wide, on pitch-pine
	2/3/2/2/	= 24	Rounded stops to ditto.
	2/2/	= 4	Fixing ¾″ bolts 8″ long, and boring pitch-pine *in scarfing.*
	2/2/	= 4	12″ × 6″ × 6″ fir cleats, framed and shaped.
	2/2/	= 4	12″ × 4″ × 6″ ditto.
	32·0 18·0	576·0	1″ sawn boarding, edges shot to roof.
	32·0 16·0	512·0	
		1088·0	
2/2/	32·0	128·0	Splayed edge to ditto.
	32·0 18·0	576·0	2½″ × ¾″ sawn battening for countess slating.
	32·0 16·0	512·0	
		1088·0	
2/	32·0	64·0	4″ × 2″ feather-edged tilter.
2/	18·0	36·0	2″ × 1″ tilting fillet *up slopes next parapets.*
2/	16·0	32·0	
		68·0	
2/3/	9·9 7·5	433·11	1″ pitch-pine (*or other material*) wrot, matched, and V-jointed diagonal boarding, in half-batten widths, nailed to soffit of rafters.
2/3/	9·9 8·0	468·0	
		901·11	9·9 7·5 17·2
2/3/	34·4 ·3	51·6	Add *for cutting and waste* 34·4
2/3/	35·6 ·3	53·3	9·9 8·0 17·9 35·6
		104·9	
2/3/4/	9·9	234·0	Bridging between rafters for boarding.
2/3/3/	9·6	171·0	3″ × 2″ pitch-pine wrot and hollow moulded angle fillet, planted on.
2/3/	9·6	57·0	5″ × 2″ ditto, but splayed at back to fit angle.
	2/2/3/4/	= 48	Mitres of 3″ × 2″ angle fillet.
2/3/3/	7·5	133·6	5″ × 2″ pitch-pine hollow moulded ceiling rib, planted on.
2/3/3/	8·0	144·0	
		277·6	

2/2/3/3/2/ 9·9	= 72	Bird's-mouthed mitres.	
3/ 9·9	29·3	1" pitch-pine wrot, matched, and V-jointed vertical boarding, in half-batten widths and 14" lengths, including bridging between rafters, one edge.	
3/ 9·9	29·3	Ditto, but with bridging between rafters, both edges.	
2/3/ 9·9	58·6	6" × 2" pitch-pine cornice, moulded and splayed (the moulding 8" girt), planted on.	
2/3/2/ 9·9	= 12	Housed ends.	
2/3/ 9·9	58·6	Splayed edge to 1" pitch-pine boarding, cross grain.	
3/ 9·9	29·3	4" × 2½" pitch-pine hollow moulded cornice on splayed ground, plugged to wall *to match plate on other side.*	
3/2/	= 6	Housed ends.	
32·0 1·3	40·0	1" gutter board and framed bearers.	
3/		2" short cross-rebated and rounded drips, in gutter.	
1/		Extra to dovetailed cesspool, 9" × 9" and 6" deep, holed for outlet.	

FLOOR AND SKIRTING
(fig. 1197)

24·0 16·0	384·0	1" flooring (*describe as in Specification*).	
10·0 3·0	30·0	Add *bay.*	
	414·0		
6·0 ·9	4·6	*chimney breast.*	
4·6 1·6	6·9	Ddt. *hearth.*	
2/½/ 2·6 2·6	6·3	*want angles of bay.*	
	17·6		
2/ 3·6	7·0	Raking, cutting, and waste on (*description*) flooring.	

5·0
2/1·9 = 3·6
———
8·6

8·6		3" × 1" glued and mitred border to hearth.	

24·0
16·0
·9
———
40·9
81·6

Bay 2/6" = 1·0
2/3·6 = 7·0
5·0
———
13·0
Less want 10·0
———
3·0
84·6

Deduct

Door A   ...   4·0
Door B   ...   4·6
Fireplace ...   4·0
———
12·6
———
72·0

72·0		11" × 1½" deal double-faced and double-moulded skirting, including splayed grounds and backings plugged to wall.	
6/		Tongued and mitred internal angles.	
4/		Ditto, obtuse.	
4/		Tongued and mitred external angles.	
4/		Housings.	
2/		Scribings to chimney-piece.	

DOOR AT A (fig. 1197)

3·0 7·0	21·0	2" deal five-panel door moulded both sides, with raised and mitred panels one side, with moulding on edge of raising.	
1/		Pair 4" (*description*) butts.	
1/		Mortise lock and furniture (*description or price*), and Preparation in door for ditto.	

3·3
2/7·1½ = 14·3
———
17·6

17·6		11" × 2" deal, cross-tongued, twice-rebated, twice-moulded, and twice-grooved linings, tongued at angles and including backings. Fixing pieces (*description*).	
8/			

4·3
2/7·7½ = 15·3
———
19·6

2/ 19·6	39·0	6½" × ¾" deal framed, splayed, and beaded grounds, rebated and tongued one edge.	

19·6
Ddt. for plinth blocks, 2/9" = 1·6
———
18·0

2/ 18·0	36·0	3" × 2" deal moulded architrave to detail, including mitres.	
2/2/	= 4	4" × 2½" deal shaped plinth blocks, 12" high, dovetailed and screwed to architrave.	
4·9 ·9 ·4	1·2	Fir lintel.	

DOOR AT B (fig. 1197)

3·6
Add for fold, 0·1
———
3·7

3·7 7·0	25·1	2" deal casement doors, each in two panels, lower panel moulded both sides, upper panel rebated, moulded one side, and open for glass, with diminished stiles.	
7·0		Hook-rebated meeting of 2" folding doors.	
7·0		1½" × ¾" moulded cover fillet to folding door.	
3·7		Splayed and throated edge to 2" casement door.	
2/		Pairs 4" (*description*) butts.	
2/		12" (*description*) brass flush bolts.	

	1/		Rebated mortise lock and furniture (*description or price*), and Preparation in door for ditto. 1·3 4·0 5·3 10·6
2/	10·6	21·0	¾″ × ½″ deal moulded fillets round glass, including mitres and fixing with brass screws and cups.
	4·3		4½″ × 3½″ oak framed, mitred, sunk, weathered, check-throated, ovolo-moulded, and grooved sill, and 1½″ × ¼″ galvanized iron tongue, bedded in white lead.
2/	9·0	18·0	4½″ × 3½″ deal framed, mitred, rebated, grooved and twice-moulded frame (A) *jambs.*
	4·3		Add *head.*
		22·3	
	4·3		4½″ × 3″ deal framed, mitred, twice-rebated, weathered, check-throated, and three times moulded transome.
	1·10		4½″ × 3″ deal framed, mitred, twice-rebated, and four times moulded mullion.
2/	1·8 1·5	4·9	2″ deal moulded fixed fanlights divided into small squares with 1″ moulded bars (in No. 2).
2/	1·8	3·4	Splayed and throated edge to 2″ casement. 4·0 2/9·0 = 18·0 22·0
	22·0		6″ × 1″ deal moulded and grooved linings, rebated and tongued one edge and at angles, and including backings. 5·0 2/9·6 = 19·0 24·0
	24·0		6½″ × ¾″ deal grounds, as last. 24·0 *Ddt. for plinth,* 2/9″ = 1·6 22·6
	22·6		3″ × 2″ deal moulded architrave, as last. 4″ × 2½″ deal plinth blocks, as last.
	2/		
	10/		Fixing pieces.
	5·6 ·9 ·4½	1·7	Fir lintel.
	3·8		Turning piece, 4½″ soffit.
			WINDOWS AT C (fig. 1198)
2/	2·9	5·6	4½″ × 3½″ oak sill as last described, but twice grooved, and 1½″ × ¼″ galvanized iron tongue, as before.
2/	2·9	5·6	4½″ × 3½″ deal frame as A *head.*
2/2/	6·0	24·0	Ditto, but water hollowed *jambs.*

2/	2·9	5·6	4½″ × 3″ deal transome, as last.
2/	2·0 4·0	16·0	2″ deal moulded hung casements in one square each (in No. 2).
2/	2·0 1·5	5·8	2″ deal moulded hung fanlight, divided into small squares, with 1″ moulded bars (in No. 2).
2/2/	2·0	8·0	Splayed and throated edge to 2″ casement. 2/1·10 = 3·8 2/3·8 = 7·4 11·0
2/	11·0	22·0	¾″ × ½″ deal moulded fillets round glass, as before.
	2/1/	= 2	Pair 3½″ (*description*) butts *to casements.*
	2/1/	= 2	Casement fastener p.c.…and fixing.
	2/1/	= 2	Casement stay bar p.c.…and fixing.
	2/1/	= 2	Pair 3″ (*description*) butts *to fanlights.*
	2/1/	= 2	Fanlight openers p.c.…and fixing. 2/5·11 = 11·10 2·8 14·6
2/	14·6	29·0	6″ × 1″ deal linings, as before. 3·8 2/6·5 = 12·10 16·6
2/	16·6	33·0	6½″ × ¾″ deal grounds, as before, and 3″ × 2″ deal architrave, as before.
2/	3·9	7·6	9½″ × 1¼″ deal cross-tongued and moulded window board, rebated and tongued on including bearers.
	2/2/	= 4	Notched and returned moulded ends.
2/	3·7	7·2	3″ × 1½″ deal, rebated and fully moulded, bed-mould tongued on, and Stopped groove in deal, and ¾″ narrow splayed grounds.
	2/2/	= 4	Returned moulded ends.
	2/8/	= 16	Fixing pieces.
2/	4·0 ·9 ·4	2·0	Fir lintel.
2/	2·3	4·6	Turning piece, 4½″ soffit.
			WINDOW AT D (fig. 1198)
	3·9 6·3	23·5	Deal-cased frames of 1″ inside and outside linings, 1½″ pulley stile tongued to linings, ½″ back lining tongued to inside lining, ½″ parting slips, ½″ parting beads, 6½″ × 3½″ oak double-sunk double-weathered twice check-throated and beaded sill, and 2″ ovolo-moulded sashes, the lower in one square, upper in small squares, with 1″ ovolo-moulded bars, double hung on (*description*) axle pulleys, (*description*) lines, iron (*lead if for plate-glass*) weights complete *in No. 1.*
	2/		Moulded horns to 2″ sash.
	1/		Sash fastener (*description or price*) and fixing.

2/		Sash lifts (*description or price*) and fixing.	4·10		7″ × 1¼″ deal moulded window board, rebated and tongued on, including bearers.
2/		Pull - down handles (*description or price*) and fixing.	2/ 3·9		Notched and returned moulded ends to 1¼″ window board.
3·0		Splay rebated meeting of 2″ sashes, and Splayed and throated edge to ditto.	4·8		Groove in oak *for window board.*
3·9		Iron tongue as before, and Groove in oak.	2/		3″ × 1½″ bed-mould as before, and Stopped groove in deal, and ¾″ narrow splayed grounds. Returned moulded ends to bed-mould.
		3·9 2/6·3 = 12·6 16·3	2/ 8/ 5·3 ·9 ·4		Fixing pieces.
16·3		Deal linings as before, but 4″ × 1″, and Groove in deal.		1·4	Fir lintel.
18·3		6½″ × ¾″ deal grounds, as before, and 3″ × 2″ deal architrave, as before.	3·0		Turning piece, 4½″ soffit.

NOTE.—In the foregoing examples it will be noted that the "dimensions" in the second column appear in varying forms. Those with one dimension are "runs" or lineal measurements, the sizes, where required, being given with the description. Those with two dimensions are "supers" or superficial measurements, the third dimension (if any) being given with the description. Those with three dimensions are "cubes". The order of writing these dimensions is for the sake of future reference and uniformity—(first) width, (second) thickness from front to back, and (third) the height. A further variation is shown in the "numbered" items which are complete in themselves. Where there are more than one item of a similar size and description, "timesing" is resorted to, *e.g.* $2/ = \times 2$, $3/2/ = 3 \times 2 = 6$. While if items are added, they are "dotted on", *e.g.* $3\cdot2/2/ = (3 + 2) \times 2 = 10$, $2/3\cdot2/2/ = 2 \times (3 + 2) \times 2 = 20$. It is usually advisable not to "times" individual items out of one set of measurements unless the repeating items form a large proportion, as this increases the difficulty of future reference.

CHAPTER II

ABSTRACTING

"Abstracting" the dimensions consists of classifying the various items in their various trades, getting them into (as nearly as possible) the order of the final bill, and collecting the items of the same denomination under their respective headings. It will be noticed that the "taker-off" takes the items as they arise without regard to the order of the bill, hence the need of some system of collection. With regard to order (and this applies also to the writing of the bill) a general rule is the following:—"Cubes", "supers", "runs", and "Nos.", with a further arrangement commencing with the "labours", then "thicknesses", starting with the least, and taking the cheapest items first. A certain amount of discretion will be necessary, as an inflexible application of these rules would by no means produce a perfect bill. It is therefore usual to classify the items somewhat, keeping, *e.g.*, the "linings", "grounds", and "architraves" together, also allowing any incidental items, to which a reference to the main items is necessary, to follow these items. This will be seen by referring to the examples of abstracting given below.

The sketches in the dimensions should, if of a slight nature, appear also on the abstract, but if they are in any way elaborate—to avoid the variation through frequent copying—a reference should be made to the column of the dimensions, *e.g.* "as sketch (col. ...)".

ESTIMATE FOR......................................

CARPENTER—I

DEAL.

Cubes.	Supers.	Runs.	Nos.
Fir in lintels and plates.	2½″ × ¾″ sawn battening for countess slating.	Splayed edge to 1″ boarding.	Fixing only ¾″ bolts, and boring for same.

Cubes. — Fir in lintels and plates.

1·2
1·7
2·0
1·4
4·2
___
10·3

Supers. — 2½″ × ¾″ sawn battening for countess slating.

1088·0

Runs. — Splayed edge to 1″ boarding.

128·0

Nos. — Fixing only ¾″ bolts, and boring for same.

2 = 1·10

2″ short cross-rebated and rounded drips in gutter.

3

Extra to dovetailed cesspool in gutter, 9″ × 9″ and 6″ deep, holed for outlet.

1

Run. — Turning piece, 4½″ soffit.

3·8
4·6
3·0
___
11·2

Fir-framed roofs.

110·11
8·9
____
119·8

Do. in short lengths.

11·6

1″ sawn boarding, edges shot to roof.

1088·0

Bridging between rafters for edge of ceiling boarding.

234·0

2″ × 1″ tilting fillet.

68·0

4″ × 2″ feather edged tilter.

64·0

Fixing pieces (description).

8
10
16
8
___
42

1″ gutter board and framed bearers

40·0

12″ × 4″ × 6″ fir cleats framed and shaped.

4   12″ × 6″ × 6″ do.

4

ESTIMATE FOR......................................

CARPENTER—II

ROOF IN PITCH-PINE (OR OTHER MATERIAL)

Cubes. — Framed in roof.

6·5   Do. in roof trusses.
18·11
____
25·4   35·11
        8·2
        45·10
        ____
        89·11

Supers. — Wrought face.

79·3
103·7
328·2
119·0
____
630·0

2½″ wrought one side and framed ribs in extra wide timbers, including scarfings (cutting measured).

59·11   3″ wrot both sides do.

59·11

Runs. — Stopped rebate 3½″ girt.

47·4   Stopped groove, 5″ girt.
114·8
____
162·0

47·1
114·8
____
162·0

Raking, cutting on 2½″ rib.

40·4   Do. on 3″ do.

40·4

Stopped chamfer, 1½″ wide.

98·6   Rounded stops.
197·0
102·0
____   68
397·6   136
        24
        ___
        228

Stopped chamfer, 3″ wide.

10·0   Square stops.
20·0
____   4
30·0   8
        ___
        12

Wrought and hollow moulded edge, 8″ girt to plate.

32·0

Circular wrought, cut, and chamfered (1½″ wide) edge to 2½″ rib.

37·0   Extra to breaks in do., as sketch.

2

Circular wrought, cut, and twice chamfered (each 1½″ wide) edge to 3″ rib.

37·0   Extra to breaks in do., as sketch.

2

Nos. — Moulded ends to 9″ × 4″ hammer beams, as sketch.

4   Do. to 9″ × 6″ do.

4

Moulded ends to 2½″ rib, including long stop to 1½″ chamfer, as sketch.

4   Do. to 3″ do., including two long stops to chamfer.

4

Fixing only ¾″ bolts and boring.

16 = 16·0
8 = 13·4
16 = 32·0
8 = 32·0
8 = 12·0
8 = 17·4
8 = 9·4
4 = 10·0
4 = 2·8
_____
80 = 144·8

Average, 1′ 10″

Hoisting and fixing roof truss, 4″ thick, 20 ft. span, and 20 ft. rise; the ridge ......feet above ground.

2   Do., but 6″ thick.

2

ESTIMATE FOR....................................,......

CARPENTER—III

ROOF in PITCH-PINE (or other Material)

Supers.	*Runs.*				*Nos.* "A"	
1″ matched and V-jointed diagonal boarding, in half-batten widths, nailed to soffit of rafters.	Splayed edge, cross-grain, on 1″ boarding.	2″ × 1½″ chamfered angle fillet, planted on.	10″ × 1½″ plinth to posts, planted on, in short lengths, and moulded 5″ girt.	5″ × 2″ twice hollow moulded ceiling rib, planted on.	2½″ triangular perforated tracery panels, 2′ 3″ × 2′ 8″ extreme, cusped and worked one side as sketch, cross-tongued as required.	Wrought on one face and edges, and framed brackets as sketch, out of 18″ × 4″ and 6 ft. long, stop-chamfered on two edges.

Supers. 1″ matched and V-jointed diagonal boarding, in half-batten widths, nailed to soffit of rafters.

901·11
104·9
——
1006·8

Runs. Splayed edge, cross-grain, on 1″ boarding.

58·6

1″ matched and V-jointed vertical boarding, in half-batten widths and 14″ lengths, including bridging between rafters to one edge.

29·3   Do., but bridging to both edges.

29·3

2″ × 1½″ chamfered angle fillet, planted on.

29·0
58·0
——
87·0

Mitred ends.

4   Ends scribed
8   over moulded plinth.
——
12

4   Scribings
8   over Moulded
——   wall
12   plate, 8″ girt.

2
4
—
6

7″ × 1½″ chamfered lining, planted on.

63·0
126·0
——
189·0

Fitted ends.

8   Do., on splay.
16
——
24   4   Irregular
8   mitre.
——
12   2
4
—
6

10″ × 1½″ plinth to posts, planted on, in short lengths, and moulded 5″ girt.

5·0
9·0
——
14·0

External mitres.

4   Fitted ends.
8
——   8
12   8
——
16

3″ × 2″ hollow moulded angle fillet, planted on.

171·0   5″ × 2″ do., splayed at back to fit angle.
——
57·0

Mitres.

48

5″ × 2″ twice hollow moulded ceiling rib, planted on.

277·6

Bird's-mouthed mitres.

72

6″ × 2″ cornice moulded and splayed (the moulding 8″ girt), planted on.

58·6   Housed ends.
——
12

4″ × 2½″ hollow moulded cornice, on splayed ground, plugged to wall.

29·3   Housed ends.
——
6

Nos. "A" 2½″ triangular perforated tracery panels, 2′ 3″ × 2′ 8″ extreme, cusped and worked one side as sketch, cross-tongued as required.

4   Do., but 2′ 6″ × 3 ft. extreme.

4

"B" Do., but 2′ 11″ × 3′ 6″ extreme and as sketch, with carved boss in centre.

4

Triangular perforated tracery panels, 2′ 3″ × 2′ 8″ as sketch "A", but 3″ thick, and cusped and worked on both sides.,

4   Do., 2′ 6″ × 3 ft. do.

4

Do., 2′ 11″ × 3′ 6″ as sketch "B" but 3″ thick, cusped and worked on both sides.

4

Wrought on one face and edges, and framed brackets as sketch, out of 18″ × 4″ and 6 ft. long, stop-chamfered on two edges.

4   Wrought on both faces and edges, and framed brackets as last sketch, but out of 18″ × 6″ and stop-chamfered on four edges.

4

ESTIMATE FOR....................................

JOINER AND IRONMONGER—I

FLOORS.

Supers.
1″ flooring (*description*).

414·0   Ddt.
17·6 Ddt.   ——
——   17·6
396·6

Runs.
Raking, cutting, and waste on 1″ ...flooring.

7·0

3″ × 1″ glued and mitred border to hearth.

8·6

SKIRTINGS

Run.
11″ × 1½″ deal double-faced and double-moulded skirting, including splayed grounds and backings plugged to wall.

72·0

Scribings to chimney piece.

2   Housings

4

Tongued and mitred internal angles.

6   Do., obtuse.

4

Tongued and mitred external angles.

4

DOORS.

Supers.
2″ deal five-panel door, moulded both sides, with raised and mitred panels one side, with moulding on edge of raising.

21·0

2″ deal casement doors, each in two panels, lower panel moulded both sides, upper panel rebated, moulded one side, and open for glass with diminished stiles.

25·1

WINDOWS.

Supers.
2″ deal moulded hung casements in one square each.

16·0 = 2

2″ deal moulded fixed fanlights, divided into small squares, with 1″ moulded bars.

4·9 = 2   2″ do., hung do.

5·8 = 2

Deal-cased frames of 1″ inside and outside linings, 1½″ pulley stile tongued to linings, ½″ back lining tongued to inside lining, ½″ parting slips, ½″ parting beads, 6½″ × 3½″ oak double-sunk, double-weathered, twice check-throated and beaded sill, and 2″ ovolo moulded sashes — lower in one square, upper in small squares, with 1″ ovolo moulded bars, double hung on ...axle pulleys, ...lines ...weights complete.

23·5 = 1

Moulded horns to 2″ sash.

2

ESTIMATE FOR ....

JOINER AND IRONMONGER—II

DOOR AND WINDOW FINISHINGS

Runs.

Groove.

16·3 Do.
stopped.

7·2
4·8
————
11·10

Splayed and throated edge to 2″ casement.

3·7 Splay-
3·4 rebated
8·0 meeting
3·0 of
———— 2″ sashes.
17·11 ————
3·0

Hook - rebated meeting of 2″ folding doors.

7·0

¾″ × ½″ deal moulded fillets, round glass, including mitres, and fixing with brass screws and cups.

21·0
22·0
————
43·0

1½″ × ¾″ moulded cover fillet to folding door.

7·0

¾″ narrow splayed grounds.

7·2
4·8
————
11·10

6½″ × ¾″ deal framed, splayed, and beaded grounds, rebated and tongued one edge.

39·0
24·0
33·0
18·3
————
114·3

6″ × 1″ deal moulded and grooved linings, rebated and tongued one edge, and at angles and backings.

22·0 4″ × 1″
29·0 do.
———— ————
51·0 16·3

11″ × 2″ deal cross-tongued, twice rebated, twice moulded, and twice grooved linings, tongued at angles, and including backings.

17·6

9½″ × 1¼″ deal cross-tongued and moulded window board, rebated and tongued on including bearers.

7·6 7″ × 1¼″
do., *not* cross-tongued.

4·10

Notched and returned moulded ends.

4
2
——
6

3″ × 1½″ deal rebated and fully moulded bed-mould, tongued on.

7·2 Returned
4·8 moulded
———— ends.
11·10
4
2
——
6

3″ × 2″ deal moulded architrave to detail, including mitres.

36·0 4″ × 2½″
22·6 shaped
33·0 plinth
18·3 blocks,
———— 12″ high,
109·9 dovetailed and screwed to architrave.

4
2
——
6

4½″ × 3″ deal framed, mitred, twice rebated, and four times moulded mullion.

1·10

4½″ × 3″ deal framed, mitred, twice rebated, weathered, check-throated, and three times moulded transome.

4·3
5·6
————
9·9

"A"
4½″ × 3½″ deal framed, mitred, rebated, grooved and twice moulded frame.

22·3 4½″ × 3½″
5·6 do., but
———— water
27·9 hollowed.

24·0

Preparation in door for mortise lock.

1
1
——
2

4½″ × 3½″ OAK framed, mitred, sunk, weathered, check-throated, ovolo moulded, and grooved sill.

4·3 4½″ × 3½″
do., but twice grooved.

5·6

Groove in oak.

3·9
3·9
————
7·6

IRONMONGERY.

Pairs.

4″ butts.

1 3½″ butts.
2
——— 2 3″
3 butts.
—— ———
2

12″ brass flush bolts.

2

Mortise lock and furniture (*description or price*).

1 Rebated,
———
1

Sash fastener (*description or price*) and fixing.

1 Sash lifts
....
and fixing.
———
2 Pull-down handles
......
and fixing.
———
2

Casement fastener, p.c. ...... and fixing.

2 Casement stay - bars p.c. ... and fixing.
———
2

Fanlight opener, p.c. .... and fixing.
———
2

1½″ × ¼″ galvanized iron tongue bedded in white lead.

4·3
5·6
3·9
————
13·6

CHAPTER III

BILLING AND PRICING

We now arrive at the final stage as far as the surveyor is concerned, *i.e.* the finished bill. The "biller" has to improve upon the work of the "abstractor" by getting the items into still better order and form, using every endeavour to make the bill consistent, and also to see that the descriptions are explanatory. If this is not the case, it is his duty to refer to the "taker-off" to rectify any weakness in this direction. It will be

noticed in the examples given, that the bill does not in all cases literally follow the abstract, either in order or description. The reason for the former is that the "abstractor" has little more than a general idea of what is to come from the dimensions beyond those he has in hand at the time, and he therefore sometimes finds that he has not room for the items in their proper positions.

ESTIMATE FOR.........................

(Date)   (Name and Address of Architect)

BILL No.......

CARPENTER

Sqs.	Ft.			(Description of materials, and any other general descriptions follow here)	£	s.	d.
	11	...	run	Turning piece, 4½″ soffit			
				FIR			
	10	...	cube	In lintels and plates, and bedding			
	120	...	"	Framed in roofs			
	12	...	"	Do. in do. in short lengths ...			
				DEAL			
10	90	...	sup.	2½″ × ¾″ sawn battening for countess slating			
10	90	...	"	1″ sawn boarding edges shot to roof			
	40	...	"	1″ gutter board and framed bearer ...			
	128	...	run	Splayed edge to 1″ boarding ...			
	234	...	"	Bridging between rafters for edge of ceiling boarding			
	68	...	"	2″ × 1″ tilting fillet			
	64	...	"	4″ × 2″ feather-edged tilter ...			
		No.	2	Fixing only ¾″ bolts 11″ long, and boring for same			
		"	2	2″ short cross-rebated and rounded drips in gutter			
		"	1	Extra to dovetailed cesspool in gutter, 9″ × 9″ and 6″ deep, holed for outlet			
		"	42	Fixing pieces (description) ...			
		"	4	12″ × 4″ × 6″ framed shaped cleats for purlins			
		"	4	12″ × 6″ × 6″ do.			
				PITCH-PINE (or other material)			
	25	...	cube	Framed in roof			
	90	...	"	Do. in roof trusses			
	630	...	sup.	Wrought face			
	60	...	"	2½″ wrought one side and framed ribs in extra wide timbers, including scarfings (cutting measured)			
	60	...	"	3″ wrought both sides do. ..			
	162	...	run	Labour stopped rebate, 3½″ girt ...			
	162	...	"	Do. stopped groove, 5″ girt ...			
	40	...	"	Do. raking, cutting on 2½″ rib ...			
	40	...	"	Do. do., on 3″ rib			
	398	...	"	Labour stopped chamfer, 1½″ wide ...			
		No.	228	Rounded stops to do.			
	30	...	run	Labour stopped chamfer, 3″ wide ...			
		No.	12	Square stops to do.			
	32	...	run	Labour wrought and hollow moulded edge, 8″ girt to plate ...			
	37	...	"	Do. circular wrought, cut, and chamfered (1½″ wide) edge to 2½″ rib ...			
		No.	2	Extra to breaks in do., as sketch ...			
	37	...	run	Labour circular wrought, cut, and twice chamfered (each 1½″ wide) edge to 3″ rib			
		No.	2	Extra to breaks in do., as last sketch			
		No.	4	Labour moulded ends to 9″ × 4″ hammer beam, as sketch ...			
		"	4	Do. do. to 9″ × 6″ do.			
		"	4	Do. moulded ends to 2½″ rib, including long stop to 1½″ chamfer, as sketch			
				Carried forward ... £			

Sqs.	Ft.				£
				Brought forward ...	£
		No.	4	Labour moulded ends to 3″ rib, including two long stops to 1½″ chamfer	
		"	80	Fixing only ¾″ bolts, average 1′ 10″ long, and boring for same ...	
		"	2	Hoisting and fixing roof trusses 4″ thick, 20 ft. span, and 20 ft. rise; the ridge ft. above ground	
		...	2	Do. do. 6″ thick, do.	
10	5		sup.	1″ matched, and V-jointed diagonal boarding, in half-batten widths, nailed to soffit of rafters	
	59	...	run	Splayed edge, cross-grain, on 1″ boarding	
	29	...	"	1″ matched and V-jointed vertical boarding, in half-batten widths and 1′ 2″ lengths, including bridging between rafters to one edge	
	29	...	"	1″ do., but including bridging to both edges	
	87	...	"	2″ × 1½″ chamfered angle fillet, planted on	
		No.	12	Mitred ends	
		"	12	Ends scribed over moulded plinth	
		"	6	Scribings over moulded wall-plate, 8″ girt	
	189	...	run	7″ × 1½″ chamfered lining, planted on	
		No.	24	Fitted ends	
		"	12	Do. on splay	
		"	6	Irregular mitres	
	14	...	run	10″ × 1½″ plinth to posts planted on in short lengths, and moulded 5″ girt	
		No.	16	Fitted ends	
		"	12	External mitres	
	171	...	run	3″ × 2″ hollow moulded angle fillet, planted on	
		No.	48	Mitres	
	57	...	run	5″ × 2″ angle fillet as last, but splayed at back to fit angle ...	
	278	...	"	5″ × 2″ twice hollow moulded ceiling rib, planted on ...	
		No.	72	Bird's mouthed mitres	
	59	...	run	6″ × 2″ cornice, moulded and splayed (the moulding 8″ girt), planted on	
		No.	12	Housed ends	
	29	...	run	4″ × 2½″ hollow moulded cornice on splayed ground, plugged to wall	
		No.	6	Housed ends	
		"	4	2½″ triangular perforated tracery panels, 2′ 3″ × 2′ 8″ extreme, cusped and worked one side as sketch, cross-tongued as required	
		"	4	2½″ do., but 2′ 6″ × 3 ft. extreme, do.	
		"	4	2½″ do., but 2′ 11″ × 3′ 6″ extreme, as sketch, with carved boss in centre	
		"	4	Triangular perforated tracery panels, 2′ 3″ × 2′ 8″, as sketch "A", but 3″ thick, and cusped and worked both sides ...	
		"	4	Do., 2′ 6″ × 3 ft., do.	
		"	4	Do., 2′ 11″ × 3′ 6″, do., but as sketch "B"	
		"	4	Wrought on one face and edges and framed brackets as sketch, out of 18″ × 4″ and 6 ft. long, stop-chamfered on two edges ...	
		"	4	Wrought on both faces and edges and framed brackets as last sketch, but out of 18″ × 6″ and do., stop-chamfered on four edges	
				Carried to Summary ...	£

ESTIMATE FOR..........................

(Date)   *(Name and Address of Architect)*

BILL No.......

JOINER AND IRONMONGER

Sqs.	Ft.			(Description of materials, and any other general descriptions follow here)	£	s.	d.
				FLOORS			
3	95	...	sup.	1" flooring (description)			
	7	...	run	Raking, cutting, and waste on 1" flooring			
	9	...	"	3" × 1" glued and mitred border to hearth			
				SKIRTINGS IN DEAL			
	72	...	"	11" × 1½" double-faced and double-moulded skirting, including splayed grounds and backings plugged to wall			
	No.	2		Scribings to chimney pieces			
	"	4		Housings			
	"	6		Tongued and mitred internal angles			
	"	4		Do. obtuse do.			
	"	4		Tongued and mitred external angles			
				DOORS IN DEAL			
	21	...	sup.	2" five-panel door, moulded both sides, with raised and mitred panels one side, with moulding on edge of raising			
	25	...	"	2" casement doors, each in two panels, lower panel moulded both sides, upper panel rebated, moulded one side, and open for glass with diminished stiles ...			
				WINDOWS IN DEAL			
	16	...	sup.	2" moulded hung casements, in one square each (in No. 2) ...			
	5	...	"	2" moulded fixed fanlights, divided into small squares, with stout moulded bars (in No. 2) ...			
	6	...	"	2" do. hung do., do. (in No. 2)			
	23	...	sup.	Cased frame of 1" inside and outside linings, 1½" pulley stile tongued to linings, ½" back lining tongued to inside lining, ½" parting slips, ½" parting beads, 6½" × 3½" *oak* double sunk, double weathered, twice check-throated and beaded sill, and 2" ovolo moulded sashes—lower in one square, upper in small squares, with 1" ovolo moulded bars double hung on...axle pulleys...lines...weights complete (in No. 1)			
	No.	2		Moulded horns to 2" sashes ...			
				DOOR AND WINDOW FINISHINGS IN DEAL			
	16	...	run	Labour groove			
	12	...	"	Do. do. stopped			
	18	...	"	Do. splayed and throated edge to 2" casement			
	3	...	"	Do. splay-rebated meeting of 2" sashes			
				Carried forward ... £			

Sqs.	Ft.			Brought forward ...	£		
7	...	run		Do. hook-rebated meeting of 2" folding doors			
43	...	"		¾" × ½" moulded fillets round glass, including mitres, and fixing with brass screws and cups ...			
7	...	"		1½" × ¾" moulded cover fillet to folding door			
12	...	"		¾" narrow splayed grounds			
114	...	"		6½" × ¾" framed, splayed, and beaded grounds, rebated and tongued one edge			
16	...	"		4" × 1" moulded and grooved linings, rebated and tongued one edge, and at angles, and including backings			
51	...	"		6" × 1" do., do.			
18	...	"		11" × 2" cross-tongued, twice rebated, twice moulded, and twice grooved linings, tongued at angles, and including backings			
5	...	"		7" × 1¼" moulded window-board, rebated and tongued on, including bearers			
8	...	"		9½" × 1¼" do., do., cross-tongued			
	No.	6		Notched and returned moulded ends			
12	...	run		3" × 1½" rebated and fully moulded bed-mould, tongued on			
	No.	6		Returned moulded ends			
110	...	run		3" × 2" moulded architrave to detail, including mitres ...			
	No.	6		4" × 2½" shaped plinth blocks, 12" high, dovetailed and screwed to architrave			
2	...	run		4½" × 3" framed, mitred, twice rebated, and four times moulded mullion			
10	...	"		4½" × 3" framed, mitred, twice rebated, weathered, check-throated, and three times moulded transome			
28	...	"		4½" × 3½" framed, mitred, rebated, grooved, and twice moulded frame			
24	...	"		4½" × 3½" do., do., and water hollowed			
	No.	2		Preparation in door for mortise lock			
				OAK			
8	...	run		Labour groove			
4	...	"		4½" × 3½" framed, mitred, sunk, weathered, check-throated, ovolo moulded, and grooved sill			
6	...	"		4½" × 3½" do., do., but twice grooved			
				IRONMONGERY *Including screws and fixing. Brass work to be fixed with brass screws.*			
	No.	2		Pairs, 3" ... butts...			
	"	2		Do. 3½"... do.			
	"	3		Do. 4" ... do.			
	"	2		12"... brass flush bolts			
	"	1		Mortise lock and furniture (description or price)			
	"	1		Rebated do.			
	"	1		Sash fastener (description or price)			
	"	2		Sash lifts (do.)			
	"	2		Pulldown handles (do.)			
	"	2		Casement fasteners, p.c. ... each ...			
	"	2		Casement stay bars, p.c. ... each ...			
	"	2		Fanlight openers, p.c. ... each			
14	...	run		1½" × ¼" galvanized iron tongue, bedded in white lead			
				Carried to Summary ... £			

NOTE.—The sketches on page 388 have not been repeated in this Bill, although they would of course be inserted in a Bill of Quantities supplied to a builder by an architect or quantity-surveyor.

The work of the Quantity Surveyor is now finished, and it remains for the Builder to insert in the Bill of Quantities his prices to enable him to make his estimate. As any example of pricing would not apply to any two districts—even if it applied to any two firms in one district—a few general principles to be observed will only be considered. In the first place there is the cost of material at the builder's yard, *i.e.* including the cost of railway carriage and cartage; secondly, the labour, and this, forming as it does such a large proportion of the cost of joiner's work, is a very important item. Note should be taken as to the quantity of any work, as if the builder possesses machinery it will be necessary for him to calculate whether the quantity in a particular item is sufficient to allow him to set the machinery for it, or whether, under the circumstances, it will pay him better to do the work by hand. Added to the cost of the prepared work are the carriage to the site and the fixing, the latter depending upon the supply and price of labour in the vicinity of the site, and whether the builder has to import labour, in which case "lodging money" has to be allowed for. And lastly, a proportion of his establishment charges, including supervision. The question of profit must also be considered.

It behoves every builder to keep a close record of the cost of his work—this only will enable him to estimate with any degree of certainty. If this was systematically done, we should not see those glaring discrepancies in tendering that are so embarrassing to architects. In short, experience is the only real teacher of this important branch of a builder's office-work, and a builder who has to rely to any extent upon the various "price books" published—useful in their way as they are—will not be likely to prove very successful in his business.

Section XV.—BUILDING LAW

BY

E. S. ROSCOE

BARRISTER-AT-LAW

AUTHOR OF "A DIGEST OF BUILDING CASES"; "A DIGEST OF THE LAW OF LIGHT".

In considering the legal rights of persons engaged in what for convenience may broadly be called building operations, it is desirable in the first place to understand clearly who these persons are. There is first of all the person who proposes to construct, alter, or repair a building—he may be the owner of the ground on which the building stands or will stand, or he may be only a tenant of a building or land. Be that as it may, he is the *Building-Owner*, which term will be used to cover all the persons who are in the position of the principal employer. Secondly, there comes the person who does the work of building or repairing, whatever may be the nature of the work; he is the *Builder*. Thirdly, we have the person who plans the work on behalf of the building-owner and for the builder to execute, and who superintends its progress; he is the *Architect*. Fourthly, we have a person who appears for a short time in most building operations—who takes out the quantities—the *Quantity Surveyor*, sometimes a separate person, sometimes merged in the architect.

These are the persons who have business relations the one with the other from the beginning to the end of the work, and the practical and logical way to understand their legal duties and liabilities is to follow out a building transaction from start to finish. Therefore the first persons who come into contact are the building-owner and the architect, and the first instructions of the building-owner are that the architect shall prepare plans and drawings. These, unless there is any distinct agreement to the contrary, are prepared for the building-owner, absolutely; in other words, they become his property just as much as a portrait painted by an artist becomes the property of the customer, subject always to the liability of the building-owner to pay to the architect a reasonable sum for his remuneration. No scale of payment—not even that prepared by the Royal Institute of British Architects—is binding on the building-owner unless agreed to by him, but this scale, if adopted by the architect, is *primâ-facie* evidence of the reasonableness of the charge. An architect may also, in his desire to obtain business, make his remuneration contingent,— for example, he may agree to make a plan of a building estate without charge, subject to his appointment as architect for the buildings to be erected thereon; then if the buildings are not erected, he is not entitled to any payment for the initial work. But it must always be borne in mind that these special cases are dangerous when regarded as precedents, since they depend so much on particular circumstances.

It is more important to point out that the architect is from the beginning to the end of the work the agent of the building-owner, and therefore stands in a fiduciary position towards him,—that is to say, it is illegal for him to take commissions or trade discounts: a trade discount given to an architect belongs to the building-owner, and the latter can recover it by legal process from the architect. It follows also from this position of principal

and agent which exists between the building-owner and the architect, that if the latter is negligent in the course of his employment, he is liable for any damages which may be thereby caused to the building. It must, however, be carefully remembered that such liability exists only when the architect is acting in the eye of the law as the building-owner's agent. When he assumes, as he has to do from time to time, the position of arbitrator, then he is not liable for negligence. On this point further observation will be found in a later portion of this work.

It also follows from this position of principal and agent, that the architect has an implied authority to pledge the building-owner's credit for such things as may be reasonably within this authority. Thus it has been held (by a jury) that, where a work was heavy or important, the architect had an implied authority to have the plans lithographed and also —being likewise the surveyor—the bills of quantities. This, however, is stated here as an example and not as a general rule.

The architect having drawn out the necessary plans and specifications, the next step (at any rate in regard to large operations) is the taking out of the quantities. This work should be done by a quantity surveyor, although it is legal for the architect who has drawn out the plans to take out the quantities also, and to be remunerated for them in addition to being paid for the plans and specifications and for the supervision of the work, if his remuneration as quantity surveyor has not been agreed to be included in his remuneration as architect. But he should be careful to point out to his employer that he is acting in this dual capacity. It is, however, decidedly more satisfactory for all parties that the quantities should be taken out by a quantity surveyor.

The quantities having been taken out, the builder will either tender for the work or a contract will be made without tenders.

But the building-owner in issuing invitations for tenders does not thereby bind himself —even if this is not stated in so many words in the advertisement—to accept any one of the tenders. The moment, however, that the tender has been sent in and has been accepted by the building-owner, there springs into existence a complete contract binding on both parties, although a more formal and detailed agreement may be subsequently drawn up.

But in tendering, a builder should exercise great care and leave a margin for mistakes. For the building-owner, when he issues advertisements for tenders, does not warrant the correctness of the plans, specifications, or quantities. Nor, on the other hand, does the quantity surveyor, in regard to the last, make any representation that they are correct. The builder must take all this on trust, and if any one of these things is incorrect, it is he who will suffer. He cannot recover damages in respect of any loss which he may have suffered in this respect, either from the building-owner, from his architect, or from the quantity surveyor.

But it may be that the contract for the work which is about to be begun is between a builder and some public body. With the creation of county councils, and urban and rural district councils and parish councils, the amount of work which has to be done by builders and contractors for corporations has undoubtedly increased of late years. It is desirable that builders should give these contracts great care. Over and over again builders have had to suffer, because of certain technicalities, by the non-payment for work to which they are entitled. The main point in regard to these contracts is that they must be under seal,—that is, the open and visible sign of the assent of the corporation must be affixed. More especially should the provisions of the Public Health Act 1875 be remembered:— Section 174 is to this effect: "Every contract made by an urban authority whereof the value or amount exceeds fifty pounds shall be in writing and sealed with the common seal of such authority"; unless this is done, whoever may be justly entitled to be paid for work done is absolutely powerless to recover a penny from such a public body. As regards public bodies not acting under the Public Health Act, the law in this respect is not so clear

as it might be, but on the whole it must be said that no contract with a corporation can be safely entered into without having the seal of such body affixed to the contract.

It may be also that the work falls through and is not continued. Then the quantity surveyor has a right to obtain payment for his work from the employer, even though, as has been said, he has been retained by the architect without special authority from the employer.

If, however, the work is carried out, the liability for the payment of the quantity surveyor is shifted on to the builder, and the quantity surveyor is entitled to claim from him the amount of his fees after the first instalment of the building price has been paid by the building-owner. Very few points of building law have been more persistently contested than this, but fortunately for all concerned it has now been settled. Ultimately, of course, the fees come out of the employer's pocket, since the builder in his tender includes a sum to cover their payment in the first instance. The usage is, it has been declared by a high legal authority, "a sensible and convenient one", and therefore it must be regarded as settled law that the builder must pay the quantity surveyor's fees.

We will therefore now assume that the work is begun, and that the builder has received the first instalment and paid the quantity surveyor. But we may take it that if the contract between the building-owner and the builder has been properly drawn up, a certificate of the architect was a condition precedent to any payment. Therefore in order to obtain the first instalment, an interim certificate should be obtained from the architect. This document should show on its face that it is a certificate, not a mere statement of figures. In other words, the document should show that it was the intention of the architect that the builder should be paid a particular sum. But being merely an interim certificate, it is subsidiary to the final certificate at the conclusion of the work. Should the architect unfortunately die in the course of the work, the contract which he has undertaken to discharge being a personal one, his orders cannot be carried out by his personal representatives, and unless his remuneration can be shown to be devisable, they will not be entitled to recover anything from the building-owner. It is obvious that architects who work together in a firm gain, therefore, some advantage over those who work singly. If, again, the architect becomes bankrupt any sums due to him must be paid to his trustee.

We will now turn to the position of the builder. He has a duty during the course of the work towards the public. This is one of some importance and nicety in towns. It may be briefly stated as a duty that he must not unreasonably inconvenience the public by his operations. Some amount of annoyance can hardly be avoided, as everyone knows, but if this annoyance becomes unreasonable, so as to cause serious and real injury to a person, for this the builder is liable in damages. It is impossible, however, either to lay down any concrete rule by which the liability of the builder can be ascertained, or to give examples which will elucidate the question, as this must depend on many attendant circumstances, such as locality, time of year, and so forth.

During the progress of a building contract it is scarcely possible to avoid some extras; in many cases they are introduced to quite an unreasonable extent, because if both building-owner and architect have clearly thought out and considered the scope and object of a proposed building, extras ought not to be numerous. Nothing, therefore, is more desirable than that an architect, before proceeding to draw plans and specifications, should point out to a building-owner the absolute necessity of clearly deciding what he wants. It may be said that an architect has only to carry out the instructions which are given to him, but so many building-owners are forgetful or ignorant of the increased cost, the delay and the general trouble caused by extras, that an architect who dwells on the importance of avoiding extras is doing valuable service to his employer and to the builder.

As a general practice it is stated in a contract that all extras must be ordered in writing, but unfortunately architects do not always bear this stipulation in mind, and even yet oftener builders are content to do extra work without a written order. No rule of law

is, however, more definitely settled than that a builder cannot recover for extras without a written order, however reasonable and just his claim for them may be. It is much better that a builder should lay himself open to the charge of being over-precise, and refuse to do any extra work without a written order, than that he should in his desire "to oblige" run the risk of serious loss at the end of the contract. If, again, the contract states that all extras shall be paid for at a price which shall be fixed by the architect,—a most important stipulation on the part of the builder,—the latter is firmly bound by it. An article may have cost him double what the architect assessed it at, but the builder is not entitled to a farthing more.

It has been said above that a builder who does extra work without a written order is running a serious risk. It is, however, a risk only, for when the final certificate comes to be made by the architect—if extras have been done without a written order, and if in this final certificate the architect states that a certain sum is due in respect of this work —then that certificate makes good the absence of the written orders. It is therefore on this final certificate that so many builders and contractors rely, when they perform extra work without a written order; and in speaking of orders for extras it must be distinctly borne in mind that as a certificate must (so to say) be a document in which an architect certifies, so an order must be a document in which an architect gives a builder authority to do certain work. A mere sketch of proposed work outside the original contract is not an authority to a builder to proceed with it.

Again, care must be taken on both sides to understand what are extras. This is why precision in specifications and drawings is so important, and why a builder should have a clear idea of what may be called the principle of the building—the extent and object of the various parts; for if he does not have this grasp of the contract, he may be under the impression that certain things are extras which are not: in other words, he may seek to make up for deficiencies in his own estimate by charging the building-owner with extras. Here again it is not desirable to give concrete examples. Each case must be considered on its merits, but practical men of sound judgment should always be able to understand when what is expressed includes or implies what is not expressed. Extras spring out of the original contract, which must not only be clearly understood in order to discover what are extras, but must also form the basis of any legal claim for work not included in it.

During the course of the work the builder is liable for damages by frost, fire, or flood, unless it is otherwise arranged in the agreement with the building user. Another important question is the liability for injuries to the person or property of third persons. Here the maxim that he who does work through another does it himself, is applicable, and so a builder or a contractor is liable *primâ facie* for injuries to the property or person of third parties. In some cases, however, the building-owner is liable, and this is a matter of much importance, because, if the building-owner is liable, the builder or contractor is under a like heavy liability. If the injured person decides to seek compensation from the building-owner, the burden on the builder is for the moment removed, but the building-owner is not responsible unless the contractor or builder fails to do something, which it would be the duty of the building-owner to do. In other words, if the work is one of an ordinary kind, the builder alone is liable if a third party's person or property is injured. On the other hand, if it is work which has an element of danger in it, then the building-owner may be liable. It may be desirable to give examples, for few points have been more keenly contested in the law-courts than this. A contractor was employed to build a bridge, and one of the workmen let fall a stone, but the building-owner was held not liable for the injury thereby caused to a third party; there was no more element of danger in building a bridge than in building a house. A contractor in another case was employed to do certain work which necessitated the under-

pinning of an adjoining land-owner's wall; when injury resulted the employer was held liable, because there was an element of danger in the work. It has, however, to be borne in mind that the liability of the building-owner does not take away the liability of the builder, because each has in the eye of the law done an injury to a third person, who has the option of obtaining damages from either of them. But the fact that a third party obtains damages from the building-owner does not merely absolve the builder from liability. Apart altogether from liability arising from legal principles, he will, in a well-drawn contract, be expressly made liable for injuries caused by the works or workmen to persons, animals, or things.

And this brings us from the progress of the work to its completion, and this question of completion is one of the most important from every point of view which arises in the course of a contract. And here it may be well to remind builders of what was said by a late distinguished judge,—that if a man does in direct terms enter into a contract to perform an impossibility, subject to a penalty, he will not be excused because it is an impossibility. This should always be borne in mind in the stress of competition, for once a contract is signed, those who have signed it are bound by it. In the particular case in which this hard saying was pronounced, the builder had by his contract bound himself to complete certain work by a specified date, and further, he had agreed to complete alterations and extras by the same date. No doubt he was under the impression that the margin of time he had allowed himself was sufficient for the original work and any extras. As a matter of fact it was insufficient. But once a contract always a contract, and having undertaken an impossibility this builder was obliged to suffer. Here, however, we have the case of a builder who by his contract had bound himself fast. Under a properly-drawn agreement he is in no such difficulty, because the broad common-sense principle of law is, that if the building-owner by his contract prevents the builder from completing the work by the date agreed on, the latter is not to suffer; so that if, for example, the builder agrees to complete a building by December 31, and in default of completion by such date to pay £2 per day as liquidated damages, and the building-owner orders extras which delay the work till the 31st of January, the proviso as to the liquidated damages comes to an end; and, speaking generally, the action of the building-owner appears to put an end to the proviso as to liquidated damages wholly. That is to say, if the extras could have been completed in two weeks over the time originally fixed for the completion of the original work, and in fact they are not completed for four weeks, the damages do not attach to the two last weeks. It may be, however, that there are words in the contract which will make them attach, and it must be said here that, if it is reasonable for liquidated damages for non-completion of work to be allowed at all, it would seem reasonable that they should come into operation on the expiration of a reasonable time for the completion of any extra work. This now is provided for in the form of building contract issued by the R.I.B.A., by which liquidated damages shall become payable at the end of the extension of time allowed by the architect for extras. This provision, while it puts an end to all difficulties, places the builder very much in the power of the architect; the date of completion in the original contract is fixed by both parties (it is a voluntary arrangement on the part of the builder), but the time for the completion of extras is assessed (so to say) by the architect.

In considering the important point of the completion of work by a builder, it must be clearly understood that the non-attachment of the clause as to liquidated damages is based on general principles. It does not apply merely to extras, but to any act of the building-owner which can be reasonably said to prevent the completion of a particular work by an agreed date. Thus, if the builder were prevented from getting to the site by the non-completion of a road to it which the building-owner was making, this would be a sufficient reason for not completing the work by the specified date

In some cases during the progress of the work the builder may not be going on with the contract with due diligence, and then, if the contract is properly drawn, the building-owner should be in a position to give notice to the builder that he must proceed with reasonable despatch. While such notice should be as precise as possible, it may be general in its terms; it must express the intention of the contract, and must give the builder fair warning that any penalties to which he may be liable for not going on with the work with due despatch will be likely to be incurred. The penalties, which usually attach to the non-proceeding of work with reasonable despatch, are the creation of a lien by the building-owner upon the builder's plant or materials, and subsequently their possession by the building-owner and their forfeiture to him. Thus there are three steps—the *notice*, the *lien*, which safeguards the building-owner, and the *forfeiture*, which practically places him in the position of the builder and (so to say) ejects the latter from the contract.

When we reach the end of the contract, there comes the important question of the final certificate. There is no part of the architect's work which demands more care than this, because he is now in the position of an arbitrator between the parties. He has now not to be a mere agent of the building-owner, as when he is drawing plans or arranging for the purchase of materials. He has to see that the employer on the one hand does not pay too much, and that the builder on the other does not receive too little. The position is very important, because it seriously affects the rights of the builder; for, being as it were a judge, the architect is not liable for negligence, and so, if he makes a mistake in the final certificate, whereby the builder is two or three hundred pounds out-of-pocket, the latter has to bear the loss; there is no legal power which can give him a remedy. What the reason of this is it is unnecessary to say, but it has been judicially stated to be because it was not thought desirable to enter into the question whether an arbitrator has been negligent or not. The question is one depending on a gradual course of judicial decisions in branches of the law not concerned with buildings. The matter has here to be looked at entirely from a practical point of view, and in this respect the first thing for a builder to bear in mind is that the final certificate of the architect cannot, even if the latter has been negligent, be contested. If a certificate is withheld, the same rule is applicable, because the certificate is only the affirmative means of showing what is due to a builder, and the absence of a certificate is a showing that nothing is due.

There is, however, an exception to the general rule that certificates cannot be set aside—in other words, that the decision of the architect is final as between the building-owner and the builder. If the certificate is in any way tainted with fraud, it can be set aside—that is to say, if the architect has purposely done a wrong act in giving a certificate, or in regard to any part of it, the document is vitiated. But again attention must be called to the fact that negligence is not fraud or wrong-doing, and that it is not easy to imagine anything more difficult to prove than that there has been fraud in regard to the giving or withholding of a certificate.

The same principles must now be cited as being applicable to the certificate of an architect as between himself and his employer. It may be admitted that, from this point of view, the question is not so simple, because the architect is the agent of the building-owner, and *primâ facie* an agent is liable to his employer for his negligence if his employer is thereby injured. But the result of the judicial decisions, taken as a whole, seems to be that the judicial character of the architect in regard to the final certificate overrules his character as agent in this respect. Certainly it would be quite illogical if he were to be in a position of judge as regards the builder, and not of judge as regards his employer. If he acts as judge towards one, he is doing the same in regard to the other, for his decision is affecting the interests of each of the two parties to the contract. Whether, however, the courts of

law have not gone too far in giving the architect a kind of judicial sanctity, is a subject of much practical importance and one well worthy of careful consideration.   From the point of view of common sense, it seems as if the judges had gone too far, and as if it would be more reasonable to make an architect liable if he is negligent in giving a certificate, for his present immunity has a tendency necessarily to make him less careful in regard to final certificates than he would otherwise be.   On the other hand, it is equally clear that the present rule of law tends to diminish litigation, to wind up contracts more quickly than would otherwise be the case, and to make builders more careful to obtain written orders for extras.   But the main point to bear in mind is, that as the law now stands the final certificate of an architect is a judicial decision,—a judicial decision, moreover, which there is no court of appeal to overrule, and therefore as final as a judgment of the House of Lords.

There is a further point, however, which must here be touched on, — namely, that though the certificate may be set aside for fraud, yet that if by the contract the decision of the architect is not to be set aside on such ground, it must stand.   It may be said that no reasonable contractor would bind himself by such a clause; this, however, is a clause which has been the subject of litigation, and which was held to prevent a builder from recovering what was due to him,—" the certificates . . . shall not be set aside or attempted to be set aside by reason, &c. . . . . or for any pretence, suggestion, charge, or insinuation of fraud, collusion, or confederacy".   It is desirable to cite these words, so that no builder may bind himself by a clause so absurdly one-sided.   Indeed, it is a question whether a builder would not be prudent to have a contract so worded that neither party should be bound by a final certificate, if it could be proved that the architect had been negligent in the making of it.   That, however, is a matter of policy, not of law.

The work is now completed, but the builder is liable if the building is deficient in any respect whether from improper materials, workmanship, or shrinkage of wood; but if the building settle from being placed on an unsuitable site, this would be due to the action of the building-owner and not of the builder.   A practical question next arises as to what is to be done with the plans and drawings.   It is obvious that they can be of no more use to the building-owner, but that they can be of great value to the architect; the natural result has been that the architect has usually retained them.   This, however, does not seem to be strictly legal, since the architect has been paid to prepare plans and to do work for his employer, so that in strict law the latter would appear to have a right to keep the plans and drawings.   As already pointed out, they can be of little or no use to him; but if the architect desires to secure his position on this point he can, having regard to the state of the law, only do so by making it a condition of his engagement that when his employment comes to an end he shall retain the plans and drawings.

In conclusion, it may be desirable to say a few words on the relation of the builder to his servant.   It has already been pointed out that under a broad legal principle a builder is answerable for the acts of his servants in regard to third parties.   The two main questions in reference to the relation of master and servant, as between a builder and his workman, are the right of the master to dismiss his servant, and his liability to him if injured in his service.

As to the first point, a builder — and these rules apply also to an architect — has a right to dismiss his workman for four causes,—namely, (a) negligence, (b) incompetence, (c) incapacity, and (d) conduct calculated to injure his master's interests.   Each of these heads raises a question of fact, and each case must be considered on its merits.   As regards (c), it would seem that even if a servant be engaged for a particular period, the builder or architect can dismiss him before the end of such period, for, as was said from the bench, a master is not bound to keep a burdensome and useless servant to the end of the term.   The same rule would apply in the case of (a) and (b).   The most difficult

cases are those which fall under head (*d*). In a case tried a good many years ago, it was
held that a builder was entitled to dismiss a workman whom he had sent to the country
to do work on a dwelling-house, upon the complaint of the building-owner that this
workman left the paths and walked about the woods disturbing the game. It may perhaps be doubted whether, at the present time, this would be considered sufficient reason for
dismissing a competent workman. But the illustration arising from actual facts is interesting and useful, since it shows how difficult it may be in some cases to decide whether certain
facts come within the rule.

So far as regards the liability of a builder for injuries to his workmen during the course
of their employment, this is now practically governed by the Workmen's Compensation Act,
1897. It may, speaking in popular language, be said that if a workman receives personal
injury by accident in the course of his employment, the employer is liable except (1) when
the workman is disabled for a shorter period than two weeks from earning full wages,
(2) when the injury is attributable to the serious and wilful misconduct of the workman,
and (3) when notice of the accident has not been given as soon as practicable after the accident and before the workman has voluntarily left his employment, or the claim for compensation is not made within six months from the occurrence of the accident, or, in case of
death, not within six months from this event. There is also an important exception which is
specially applicable to builders, namely, that the Act does not apply to workmen engaged on
a building less than 30 feet in height—that is, in fact, at the moment of the accident, and not
in plan or design,—unless machinery is being employed on it, in which case the building
may be less than 30 feet in height. It must also be in course of construction or reparation
by means of a scaffolding, or being demolished. As to what constitutes scaffolding, this
is a question partly of fact and partly of law, and it has been now definitely decided by
the House of Lords that the word includes any erection which would popularly be termed
a scaffolding.

An employer is also liable even if he makes a sub-contract for part of the work, but
such contracts must be in respect of the main work itself and not in respect of something
auxiliary or incidental to it.

It is also necessary, in order to enable a workman to recover damages, that he must be
employed on, in, or about a building; whether he is "on, in, or about" is a question of fact for
the judge, but, having regard to recent decisions in the House of Lords, the largest possible
construction should be given to these words. It has been held, for example, that a workman injured in loading a cart close to the entrance of a building, came within the words.
Again, the words "constructed or repaired" should not be limited, and they now appear to
cover almost every operation connected with a building.

APPENDIX

MAXIMUM WORKING LOADS OF TIMBER BEAMS
SUPPORTED AT THE ENDS AND UNIFORMLY LOADED (SEE NOTES)

Clear Span. Breadth in.	Depth in.	4 ft. cwt.	6 ft. cwt.	8 ft. cwt.	10 ft. cwt.	12 ft. cwt.	14 ft. cwt.	16 ft. cwt.	18 ft. cwt.	20 ft. cwt.	22 ft. cwt.	24 ft. cwt.
1	3	1·48	·99	·74	...	...	...	...	...	...	...	...
1	3½	2·02	1·35	1·01	...	...	...	...	...	...	...	...
1	4	2·64	1·76	1·32	1·05	...	...	...	...	...	...	...
1	4½	3·34	2·23	1·67	1·33	...	...	...	...	..	...	...
1	5	4·13	2·75	2·06	1·65	1·37	..	...	...	...	...	...
1	5½	5·00	3·33	2·50	2·00	1·66	...	...	...	...	...	...
1	6	5·95	3·96	2·97	2·38	1·99	1·70	...	...	...	...	...
1	7	8·10	5·40	4·05	3·24	2·70	2·31	2·02	...	...	...	...
1	8	10·58	7·05	5·29	4·23	3·52	3·02	2·64	2·35	...	...	...
1	9	13·39	8·92	6·69	5·35	4·46	3·82	3·34	2·97	2·67	...	...
1	10	16·53	11·02	8·26	6·61	5·51	4·72	4·13	3·67	3·30	3·00	...
1	11	20·00	13·33	10·00	8·00	6·66	5·71	5·00	4·44	4·00	3·63	3·33
1	12	23·80	15·87	11·90	9·52	7·93	6·80	5·95	5·29	4·76	4·32	3·96
1	14	32·40	21·60	16·20	12·96	10·80	9·25	8·10	7·20	6·48	5·89	5·40
1	16	42·32	28·21	21·16	16·93	14·10	12·09	10·58	9·40	8·46	7·69	7·05
1	18	53·82	35·71	26·78	21·42	17·85	15·30	13·39	11·90	10·71	9·74	8·92
1	20	61·13	44·09	33·06	26·45	22·04	18·89	16·53	14·69	13·22	12·02	11·02

NOTES

The figures in this table apply to beams of American white pine (known in this country as "yellow pine"), American spruce, and Baltic whitewood, the value of the coefficient of rupture (*f*) being taken at 4000, and the factor of safety at 6.

For *red or yellow deal* and *sequoia*, add one-eighth to the loads given in the table.

For *American red pine, cedar,* and *cypress*, add one-fourth to the loads given in the table.

For *oak* and *pitch-pine* (shortleaf), add one-half to the loads given in the table.

For *teak* and *pitch-pine* (longleaf), add three-fourths to the loads given in the table.

For *jarrah* and *karri*, multiply by 2¼ the loads given in the table.

For central loads, divide the figures in the table by 2. For beams of greater breadth than 1 inch, multiply the safe load given in the table by the breadth in inches; thus, to find the safe load for a joist 3 in. broad and 7 in. deep, with a clear span of 12 ft : find in the table the safe load for a joist 7 in. deep and 12 ft. span, namely 2·70 cwt.; multiply this by 3 (the breadth of the joist for which the safe load is required), and we obtain the answer, 8·10 cwt. as the safe load for whitewood; for red deal add one-eighth, thus, 8·10 + 1·01 = 9·11 cwt.

N.B.—The figures are calculated from the formula for *strength* and not from that for *deflection*. In using the table it is therefore necessary to select a section deep in proportion to the breadth. As a rough rule, it may be said that the depth of a timber beam ought not to be less than one-fifteenth of the span.

SCANTLINGS FOR FLOORS, ROOFS, &c.
APPROVED BY THE CORPORATION OF WEST HAM AND SOME OTHER LOCAL AUTHORITIES

COMMON FLOOR-JOISTS NOT MORE THAN 15 INCHES FROM CENTRE TO CENTRE

DOMESTIC BUILDINGS.					BUILDINGS OF THE WAREHOUSE CLASS.			
Clear Span.		Breadth.	Depth.	Rows of Bridging.	Clear Span.		Breadth.	Depth.
ft. in.	ft. in.	in.	in.	No.	ft. in.	ft. in.	in.	in.
up to	3 4	2½	3	0	up to	3 0	3	4½
3 5 to	5 4	2	4	0	3 1 to	4 0	2½	6
5 5 "	7 4	2½	4½	0	4 1 "	5 0	2½	7
7 5 "	9 4	2½	5	0	5 1 "	6 0	3	7
9 5 "	11 4	2½	6	1	6 1 "	7 0	3	7½
11 5 "	13 4	2½	7	1	7 1 "	8 0	3	8
13 5 "	14 4	3	7	1	8 1 "	10 0	3	9
14 5 "	16 4	3	8	1	10 1 "	12 0	3	10
16 5 "	18 4	3	9	2	12 1 "	14 0	3	11
18 5 "	20 4	3	10	2	14 1 "	16 0	3	12
20 5 "	22 4	3	11	2	16 1 "	18 0	6	9
......		...	...	...	18 1 "	20 0	4½	11

FLOOR-BEAMS NOT MORE THAN 10 FEET FROM CENTRE TO CENTRE

DOMESTIC BUILDINGS.				BUILDINGS OF THE WAREHOUSE CLASS.	
Span.		Breadth.	Depth.	Breadth.	Depth.
ft. in.	ft. in.	in.	in.	in.	in.
8 0 to	10 0	6	9	11	12
10 1 "	12 0	6	11	12	13
12 1 "	14 0	9	11	13	14
14 1 "	16 0	9	13	14	15
16 1 "	18 0	10	14	15	18
18 1 "	20 0	11	15	15	24
......		...	...	...	...

COMMON RAFTERS.

"Clear Bearing."		Breadth.	Depth.
ft. in.	ft. in.	in.	in.
up to	7 6	2	4
7 7 to	9 0	2	5

PURLINS.

"Clear Bearing."		Distance Apart.		Breadth.	Depth.
ft. in.	ft. in.	ft. in.	ft. in.	in.	in.
up to	7 6	up to	9 0	2	7
7 7 to	8 4	"	9 0	3	7
8 5 "	10 4	"	9 0	3	9
10 5 "	12 4	"	6 0	6	8
" "	"	6 1 to	7 6	6	8½
" "	"	7 7 "	9 0	6	9
12 4 "	14 4	up to	6 0	6	9
" "	"	6 1 to	7 6	6	9½
" "	"	7 7 "	9 0	6	10
14 4 "	16 4	up to	6 0	6	11
" "	"	6 1 to	7 6	6	11½
" "	"	7 7 "	9 0	6	12
16 4 "	18 4	up to	6 0	7	11
" "	"	6 1 to	7 6	7	11½
" "	"	7 7 "	9 0	7	12

NOTES

Trimmers and trimming-joists must be one inch greater *in breadth and depth* than the common joists used in the floor, and no trimmer must carry more than six common joists.

In all cases the figures are for "any species of fir or pine of sound and good quality" Instead of the scantlings given in the tables, others may be substituted, but the breadth multiplied by the square of the depth of the substituted timber must be not less than the *bd²* of the timber in the table.

MAXIMUM SPANS OF FLOOR-JOISTS

Supported at the Ends and Uniformly Loaded (see Notes)

I.—FOR HOUSES

Estimated Load, $\frac{5}{8}$ cwt. per sq. ft.

Bdth in.	Dpth in.	A ft. in.	B ft. in.	C ft. in.	D ft. in.
2	4	5 2	5 6	6 4	6 10
2	5	6 6	6 10	7 11	8 7
2	6	7 9	8 3	9 6	10 3
2	7	9 1	9 7	11 1	12 0
2	8	10 4	11 0	12 8	13 8
2	9	11 8	12 4	14 3	15 5
2	10	13 0	13 9	15 11	17 2
2	11	14 3	15 1	17 6	18 11
2	12	15 7	16 6	19 1	20 7
2½	4	5 9	6 2	7 1	7 8
2½	5	7 3	7 8	8 10	9 7
2½	6	8 8	9 3	10 8	11 6
2½	7	10 2	10 9	12 5	13 5
2½	8	11 7	12 4	14 2	15 4
2½	9	13 1	13 10	15 11	17 3
2½	10	14 6	15 5	17 9	19 2
2½	11	16 0	16 11	19 6	21 1
2½	12	17 5	18 6	21 4	23 1
3	4	6 4	6 9	7 9	8 5
3	5	7 11	8 5	9 9	10 6
3	6	9 6	10 1	11 8	12 7
3	7	11 1	11 9	13 8	14 8
3	8	12 9	13 6	15 7	16 10
3	9	14 4	15 2	17 7	18 11
3	10	15 11	16 10	19 6	21 0
3	11	17 6	18 6	21 6	23 1
3	12	19 1	20 3	23 5	25 3
4	4	7 4	7 9	9 0	9 8
4	5	9 2	9 9	11 3	12 2
4	6	11 0	11 8	13 6	14 7
4	7	12 10	13 8	15 9	17 0
4	8	14 8	15 7	18 0	19 5
4	9	16 6	17 7	20 3	21 10
4	10	18 4	19 6	22 6	24 4
4	11	20 2	21 6	24 9	26 9
4	12	22 0	23 5	27 0	29 2

II.—FOR HOUSES OF BETTER CLASS, SCHOOLS, &c.

Estimated Load, $\frac{3}{4}$ cwt. per sq. ft.

Bdth in.	Dpth in.	A ft. in.	B ft. in.	C ft. in.	D ft. in.
2	4	4 9	5 0	5 9	6 3
2	5	5 11	6 3	7 3	7 10
2	6	7 1	7 6	8 8	9 5
2	7	8 3	8 9	10 2	11 0
2	8	9 6	10 0	11 7	12 6
2	9	10 8	11 3	13 1	14 1
2	10	11 10	12 7	14 6	15 8
2	11	13 0	13 10	16 0	17 3
2	12	14 3	15 1	17 5	18 10
2½	4	5 3	5 7	6 6	7 0
2½	5	6 7	7 0	8 1	8 9
2½	6	7 11	8 5	9 9	10 6
2½	7	9 3	9 10	11 4	12 3
2½	8	10 7	11 3	13 0	14 0
2½	9	11 11	12 8	14 7	15 9
2½	10	13 3	14 1	16 3	17 6
2½	11	14 7	15 5	17 10	19 3
2½	12	15 11	16 10	19 6	21 0
3	4	5 9	6 2	7 1	7 8
3	5	7 3	7 8	8 10	9 7
3	6	8 8	9 3	10 8	11 6
3	7	10 2	10 9	12 5	13 5
3	8	11 7	12 4	14 3	15 4
3	9	13 1	13 10	16 0	17 3
3	10	14 6	15 5	17 9	19 2
3	11	16 0	16 11	19 6	21 1
3	12	17 5	18 6	21 4	23 1
4	4	6 8	7 1	8 2	8 10
4	5	8 4	8 10	10 3	11 1
4	6	10 0	10 8	12 4	13 4
4	7	11 8	12 5	14 4	15 7
4	8	13 5	14 3	16 5	17 9
4	9	15 1	16 0	18 6	20 0
4	10	16 9	17 9	20 6	22 2
4	11	18 5	19 6	22 7	24 5
4	12	20 1	21 4	24 8	26 8

III.—FOR SCHOOLS, OFFICES, CHURCHES AND CHAPELS, &c.

Estimated Load, 1 cwt. per sq. ft.

Bdth in.	Dpth in.	A ft. in.	B ft. in.	C ft. in.	D ft. in.
2	4	4 1	4 4	5 0	5 5
2	5	5 1	5 5	6 3	6 9
2	6	6 2	6 6	7 6	8 1
2	7	7 2	7 7	8 9	9 4
2	8	8 2	8 8	10 0	10 10
2	9	9 3	9 9	11 3	12 2
2	10	10 3	10 10	12 7	13 7
2	11	11 3	12 0	13 10	14 11
2	12	12 4	13 1	15 1	16 3
2½	4	4 7	4 10	5 7	6 1
2½	5	5 9	6 1	7 0	7 7
2½	6	6 10	7 3	8 5	9 1
2½	7	8 0	8 6	9 10	10 7
2½	8	9 2	9 9	11 3	12 2
2½	9	10 4	10 11	12 8	13 8
2½	10	11 6	12 2	14 0	15 2
2½	11	12 7	13 4	15 5	16 8
2½	12	13 9	14 7	16 10	18 3
3	4	5 0	5 4	6 2	6 8
3	5	6 3	6 8	7 8	8 4
3	6	7 6	8 0	9 3	10 0
3	7	8 9	9 4	10 9	11 8
3	8	10 0	10 8	12 4	13 4
3	9	11 3	12 0	13 10	15 0
3	10	12 7	13 4	15 5	16 8
3	11	13 10	14 8	16 11	18 4
3	12	15 1	16 0	18 6	20 0
4	4	5 9	6 2	7 1	7 8
4	5	7 3	7 8	8 10	9 7
4	6	8 8	9 3	10 8	11 6
4	7	10 2	10 9	12 5	13 5
4	8	11 7	12 4	14 2	15 4
4	9	13 1	13 10	16 0	17 3
4	10	14 6	15 5	17 9	19 2
4	11	16 0	16 11	19 7	21 1
4	12	17 5	18 6	21 4	23 1

IV.—FOR SHOPS, WAREHOUSES, &c.

Estimated Load, $1\frac{1}{2}$ cwt. per sq. ft.

Bdth in.	Dpth in.	A ft. in.	B ft. in.	C ft. in.	D ft. in.
2	6	5 0	5 4	6 2	6 8
2	7	5 10	6 2	7 2	7 9
2	8	6 8	7 1	8 2	8 10
2	9	7 6	8 0	9 3	10 0
2	10	8 4	8 10	10 3	11 1
2	11	9 2	9 9	11 3	12 2
2	12	10 0	10 8	12 4	13 4
2½	6	5 7	5 11	6 10	7 5
2½	7	6 6	6 11	8 0	8 8
2½	8	7 6	7 11	9 2	9 11
2½	9	8 5	8 11	10 4	11 2
2½	10	9 4	9 11	11 6	12 5
2½	11	10 3	10 11	12 7	13 8
2½	12	11 3	11 11	13 9	14 11
3	6	6 2	6 6	7 6	8 1
3	7	7 2	7 7	8 9	9 6
3	8	8 2	8 8	10 0	10 10
3	9	9 3	9 9	11 4	12 2
3	10	10 3	10 10	12 7	13 7
3	11	11 3	11 11	13 10	14 11
3	12	12 4	13 1	15 1	16 3
4	6	7 1	7 6	8 8	9 5
4	7	8 3	8 9	10 2	11 0
4	8	9 6	10 0	11 7	12 6
4	9	10 8	11 4	13 1	14 1
4	10	11 10	12 7	14 6	15 8
4	11	13 0	13 10	16 0	17 3
4	12	14 3	15 1	17 5	18 10

V.—FOR WAREHOUSES, WORKSHOPS CONTAINING HEAVY MACHINERY, &c.

Estimated Load, 2 cwt. per sq. ft.

Bdth in.	Dpth in.	A ft. in.	B ft. in.	C ft. in.	D ft. in.
2	6	4 4	4 7	5 4	5 9
2	7	5 1	5 4	6 2	6 8
2	8	5 9	6 2	7 1	7 8
2	9	6 6	6 11	8 0	8 7
2	10	7 3	7 8	8 10	9 7
2	11	8 0	8 5	9 9	10 7
2	12	8 8	9 3	10 8	11 6
2½	6	4 10	5 2	5 11	6 5
2½	7	5 8	6 0	6 11	7 6
2½	8	6 6	6 10	7 11	8 7
2½	9	7 3	7 9	8 11	9 8
2½	10	8 1	8 7	9 11	10 9
2½	11	8 11	9 5	10 11	11 9
2½	12	9 9	10 4	11 11	12 10
3	6	5 4	5 8	6 6	7 0
3	7	6 2	6 7	7 7	8 3
3	8	7 1	7 6	8 8	9 5
3	9	8 0	8 6	9 9	10 7
3	10	8 10	9 5	10 10	11 9
3	11	9 9	10 4	11 11	12 11
3	12	10 8	11 4	13 1	14 1
4	6	6 2	6 6	7 6	8 1
4	7	7 2	7 7	8 9	9 6
4	8	8 2	8 8	10 0	10 10
4	9	9 3	9 9	11 4	12 2
4	10	10 3	10 10	12 7	13 7
4	11	11 3	11 11	13 10	14 11
4	12	12 4	13 1	15 1	16 3

NOTES

In all cases the joists are assumed to be 15 inches from centre to centre, and a factor of safety of 6 is allowed.

Column A.—American white pine, spruce, and Baltic whitewood ($f = 4000$).

Column B.—Red or yellow deal ($f = 4500$).

Column C.—Oak and shortleaf "pitch-pine" ($f = 6000$).

Column D.—Teak and longleaf "pitch-pine" ($f = 7000$).

N.B.—The figures are calculated from the formula for *strength*, and not from that for *deflection*. In using the table it is therefore necessary to select a section deep in proportion to the breadth. As a rough rule, it may be said that where plastered ceilings are supported by the joists, the depth of the joists in Classes I and II ought not to be less than $\frac{1}{17}$ of the span for Column A, $\frac{1}{18}$ for Column B, $\frac{1}{20}$ for Column C, and $\frac{1}{21}$ for Column D, and in Classes IV and V, $\frac{1}{15}$, $\frac{1}{16}$, $\frac{1}{18}$, and $\frac{1}{19}$ for A, B, C, and D respectively.

GLOSSARY

OF TERMS USED IN ARCHITECTURE AND BUILDING

ABACISCUS.—1. Any flat member.—2. The square compartment of a mosaic pavement.

ABACUS.—A table constituting the upper or crowning member of a column and its capital. It is rectangular in the Tuscan and Doric orders; but in the Corinthian and Composite orders its sides are curved inwards. These curves are called the *arches* of the abacus, and the meeting of the curves its *horns*. The abacus in the Grecian Doric order is a square member. In the Tuscan and Roman Doric it has a moulding and fillet. In the Grecian Ionic the profile of its side is an ovolo or ogee, and in the Roman Ionic an ovolo or ogee with a fillet over; in the Corinthian and Composite orders its mouldings are a cavetto, a fillet, and an ovolo. In mediæval architecture the abacus is strongly marked in the earlier styles, but loses this character in the later styles.

Corinthian Abacus.

ABSCISSA.—A part of the diameter, or transverse axis of a conic section, intercepted between the vertex, or some other fixed point, and a semi-ordinate. Thus, in the parabolic figure B C A, the part of the axis D C intercepted between the semi-ordinate B D, and the vertex C, is an abscissa.

ABUT, TO.—To adjoin, to be contiguous to; generally contracted to *Butt*, as in "*Butt*"-joints, these being plain joints formed by one piece abutting against another.

ABUTMENT.—1. The solid pier or mound of earth from which an arch springs.—2. *Abutments of a bridge*, the solid extremities on or against which the arches rest.

ACANTHUS.—The plant bear's-breech, the leaves of which are imitated in the foliage of the Corinthian and Composite capitals.

Acanthus.

ACROTER, Acroterium, Acroteria.—A small pedestal, placed on the apex or angles of a pediment, for the support of a statue or other ornament. The term is also used to denote the pinnacles or other ornaments on the horizontal copings or parapets of buildings, and which are sometimes called *acroterial ornaments*.

ADZE (formerly written *Addice*). —A cutting tool used for chopping a surface of timber. It consists of a blade of iron, forming a portion of a cylindrical surface, ground to to an edge from the concave side outwards at one end, and having a socket at the other end for the handle, which is set radially. The handle is from 24 to 30 inches long. The weight of the blade is from 2 to 4 lbs.

Pediment with Acroteria, A A A.

AISLE (pronounced *Ile*).—The wing of a building; usually applied to the lateral divisions of a church, which are separated from the central part, called the nave and choir, by pillars and piers.

A LA GRECQUE, A LA GREC.—One of the varieties of the fret ornament.

A la Grecque.—Greek Border Ornament.

ALBURNUM.—The white and softer part of the wood of exogenous plants; sap-wood.

ALCOVE.—A recessed part of a room.

AMBO.—A pulpit or reading-desk in early Christian churches.

AMBRY.—1. A cupboard or closet.—2. In ancient churches a cupboard formed in a recess in the wall, with a door to it, placed by the side of the altar, to contain the sacred utensils.

ANGLE-BAR.—The vertical bar at each angle of windows, sky-lights, &c., constructed on a polygonal plan.

ANGLE-BEAD, Angle-Staff.—A piece of wood fixed vertically upon the exterior or salient angle of an apartment, to preserve it from injury, and also to serve as a guide by which to float the plaster. It is called also *staff-bead*.

ANGLE-BRACE.—1. A piece of timber fastened at each end to one of the pieces forming the adjacent sides of a system of framing, and subtending the angle formed by their junction. When it is fixed between the opposite angles of a quadrangular frame it is called a *diagonal brace*. It is also called *angle-tie* and *diagonal-tie*.—2. A boring tool for working in corners and other places where there is not room to swing round the cranked handle of the ordinary brace. It is made of metal, with a pair of bevel pinions, and a winch handle, which revolves at right angles to the axis of the hole to be bored.

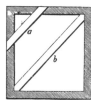
a, Angle-Brace. b, D.a-gonal Brace.

ANGLE-BRACKET.—A bracket in an interior or exterior angle, and not at right angles with the planes which form it.

ANGLE-CAPITAL.—An Ionic capital on the flank column of a portico, the exterior volute being placed at an angle of 135° with the plane of the frieze, on front and flank.

ANGLE-IRON.—Sometimes known as L-iron, from the shape of its cross-section.

ANGLE-RAFTER.—See HIP.

ANGLE-RIB.—A curved piece of timber placed in the angle between two adjacent sides of a coved or arched ceiling, so as to range with the common ribs.

ANNULAR VAULT.—A vault springing from two walls, both circular on plan, the one being concentric to the other.

ANNULATED COLUMNS.—Columns clustered together and joined by rings or bands. They are much used in First Pointed architecture.

ANNULET.—A small moulding, whose horizontal section is circular. It is used indiscriminately as a synonym for list, cincture, fillet, tenia, &c. Correctly, annulets are the fillets or bands which encircle the lower part of the Doric capital, above its neck or trachelium.

ANTÆ.—The pier-formed ends of the side walls of temples, when they are prolonged beyond the face of the end walls. A term applied to pilasters when they stand opposite a column. A portico *in antis* is one in which columns stand between antæ.

Portico in Antis. A A, Antæ.

ANTEFIXÆ. — Upright blocks ornamented on the face, placed at regular intervals on the crowning member of a cornice. These ornaments were originally used to terminate the ends of the covering tiles of the roof.

ANTHEMION.—A floral ornament common in Greek and Roman architecture, known also as the Greek honeysuckle.

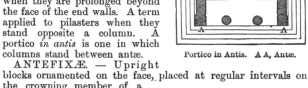
Antefixæ.

APART. — A term used to denote the clear distance or space between two objects. It must not be confused with the term "from centre to centre". Thus, if joists 2 inches thick are fixed 14 inches from centre to centre, they will be 12 inches apart.

APEX.—The highest point or summit, usually applied to the point of a gable.

APOPHYGEE, Apollusis, Apophysis.—The parts at the top and bottom of the shaft of a column which spring out to meet the edges of the fillets, usually in a concave sweep or cavetto, and often called the *spring* or *scape.*

APRON.—1. A platform or flooring of plank at the entrance of a lock, on which the gates are shut.—2. A term used by plumbers for a straight sheet of lead cover-flashing, in distinction to "step" flashing.

APRON-LINING.—The facing of the apron-piece.

APRON-PIECE.—A piece of timber fixed into the walls of a staircase, and projecting horizontally, to support the carriage pieces and joisting in the half-spaces or landings. It is called also *pitching-piece.*

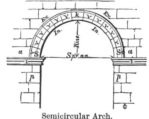
Arabesque.

APSE.—A term applied to that part of any building which has a circular or polygonal termination. The eastern portion of the church, where the clergy sat and where the altar was placed.

ARABESQUE.—A style of ornament composed of representations of fruit and flowers and other objects interwoven together. In pure Mahomedan arabesques no animal representations are used.

ARCADE.—A series of arches supported on piers or pillars, used generally as the screen and roof support of an ambulatory or walk; but in the architecture of the Middle Ages also applied as an ornamental dressing to a wall, as in the figure.

ARCH.—A structure composed of separate inelastic bodies, having the shape of truncated wedges, so as to retain their position by mutual pressure. The names of the different parts are given in the illustration of a semicircular arch. Arches are designated in two ways: first, in a general manner, according to their properties, uses, position in a building, or their exclusive employment in a particular style of architecture; second, specifically, according to the curve of their intrados, as circular, segmental, elliptical, or flat; or from the resemblance of the whole contour of the curve to some familiar object, as lancet-arch and horse-shoe; or from the method used in describing the curve, as three-centred, four-centred, and the like. When any arch has one of its imposts higher than the other it is said to be *rampant.*

Arcade, Romsey Church, Hampshire.

Semicircular Arch.

a a, Abutments.   *S,* Springers. *vv,* Voussoirs. *ii,* Imposts. *In.,* Intrados. *k,* Keystone. *p p,* Piers. *Ex.,* Extrados.

ARCHITRAVE.—1. The lower division of an entablature, or that part which rests immediately on the capital of the column. It is sometimes called the *epistylium.* See Column.—2. The plain or moulded lining on the faces of the jambs and lintels of a door or window opening, or niche.

ARCHITRAVE-CORNICE.—An entablature consisting of an architrave and cornice only, the frieze being omitted.

ARCHIVOLT.—The architrave or band of mouldings on the face of an arch following the contour of the intrados.

AREA.—The superficial content of any figure. Also a small open space in front of windows, &c., below the level of the ground.

ARRIS.—The line in which two straight or curved surfaces of any body, forming an exterior angle, meet each other; an edge.

ARRIS-FILLET.—A triangular piece of wood used to raise the slates of a roof when they abut against the shaft of a chimney or a wall, so as to throw off more effectually the rain from the joining. It is called also a *tilting-fillet.*

ARRIS-GUTTER.—A wooden gutter of the form of a V in section, fixed to the eaves of a building.

ASBESTOS.—An incombustible mineral of a fibrous nature, used for packing wood floors, partitions, &c., and also in the manufacture of fire-resisting paint.

ASHLERING.—Timber quarterings in garrets for affixing lath to, in forming partitions, to cut off the acute angle made by

the meeting of the sloping roof with the floor. They are usually perpendicular to the floor, and fixed at the top to the rafters.

ASPHALT. — A natural bituminous substance used for bedding wood-block floors, &c. It is impervious to water. "British Asphalt" is an inferior preparation made by boiling coal-tar and pitch together with fine sand, slaked lime, &c.

ASTRAGAL. — A small moulding, semicircular in its profile. It is frequently ornamented by being carved into the representation of beads or berries.

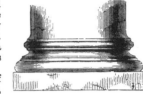
Atlantes, in the Baths, Pompeii.

ATLANTES.—A term applied to figures [or half-figures of men used in the place of columns or pilasters, to support an entablature. They are called also *Telamones.*

ATTACHED COLUMNS.—Those which project three-fourths of their diameter from the wall.

ATTIC BASE. — A moulded base used by the ancients in the Ionic order or column, and by Palladio and others in the Doric. It consists of an upper torus, a scotia, and lower torus, with fillets between them.

ATTIC ORDER.—An order of small square pillars, or pilasters, above the main cornice of a building.

Attic Base.

AUGER.—A tool used for boring large holes in wood. It consists of an iron blade ending in a steel bit, and having a handle placed at right angles to the blade.

AWL.—An iron instrument for piercing small holes. See Brad and Brad-Awl.

AXIS.—1. In geometry, the straight line in a plane figure, round which it revolves to generate a solid.—2. Generally, a supposed right line drawn from the centre of one end to the centre of the other, in any figure.

AXLE-PULLEY.—A sash-pulley; a pulley over which the cord or line connecting a window-sash to one of its balance-weights is passed.

B

BACK.—The side opposite the face or breast. When a piece of timber is laid in a horizontal or an inclined position the under side is called the breast, and the upper side the *back.* Thus, we have the back of a handrail, the back of a rafter, &c., meaning the upper side.—*Back-flaps,* those parts of a folding window-shutter which are concealed when the shutter is folded into the shutter-box or casing.—*Back-lining,* the piece of a sash-frame parallel to the pulley piece and next to the jamb.— *Window-back,* the space between the sill of a window and the floor, covered with boarding or panelling; the reveals are known as *elbows.*

BACKING.—The angle to which the back or upper surface of a hip-rafter must be cut.

BALCONY.—A frame of wood, iron, or stone projecting from an upper story of a building, and supported by pillars, columns, or consoles, and encompassed with a balustrade, railing, or parapet.

BALDACHIN, Baldachino. — A canopy over an altar, bishop's throne, &c.

BALK.—A squared log or large piece of timber.

BALL-FLOWER.—An ornament resembling a ball enclosed in a circular flower, the three petals of which form a cup round it. The ball-flower ornament is usually found inserted in a hollow moulding, and may be considered as one of the characteristics of the Decorated or Second Pointed style.

BALUSTER.—A small moulded column used in balustrades. In joinery, a plain or moulded member of wood or iron resting on the outer-string or tread of a flight of stairs, or on the floor of a landing, &c., and supporting a handrail. Also any similar member. The lateral part of the Ionic capital.

Ball-Flower.

BALUSTRADE.—A row of balusters set on a continuous plinth, and surmounted by a cap or rail, serving as a fence for altars, balconies, terraces, steps, staircases, tops of buildings, &c.

BAND.—In classic architecture, any flat member with small projection. In mediæval architecture, the suite of mouldings which girds the middle of the shafts in the Early English style.

Balustrade.

BANDING-PLANE.—A plane intended for cutting out grooves and inlaying strings and bands in straight and circular work.

BAND-SAW.—A machine with a fine endless saw passing over two pulleys; also the saw itself.

BAR.—A sash-bar; a kind of fastening for doors.

BAR, or BARRED DOOR.—The Scottish synonym for batten or ledged door; a door formed of narrow deals joined by grooving and tonguing or by rebating, and secured by bars or ledges nailed across the back.

BARGE-BOARDS, called also GABLE-BOARDS.—The raking-boards at the gable of a building, placed to cover the ends of the roof timbers when they project beyond the walls. They are sometimes called verge-boards.—Barge-Couples, the exterior couples of a roof which project beyond the gable.—Barge-Course, the course of tiles which covers and overhangs the gable-wall of a building, and is made up below with mortar; also, a coping to a wall formed of a course of bricks set on edge.

BAR-POSTS.—Posts driven into the ground to form the two sides of a field gateway. They have holes corresponding to each other, into which bars are inserted to form the fence.

BASE.—The bottom of anything, considered as its support or that whereon it stands or rests. The base of a pillar or column is the moulding or series of mouldings between the bottom of the shaft and the pedestal or plinth; and in the Grecian Doric, the steps on which the column stands form its base. The lowest part of a pedestal, and the plain or moulded fittings which surround the bottom of a wall next the floor.

BASEMENT.—1. The ground-floor on which the order or columns which decorate the principal story are placed.—2. A story below, or partly below, the level of the street.

BATTEN.—A piece of timber from $1\frac{1}{2}$ inch to 7 inches broad, and $\frac{1}{2}$ inch to $2\frac{1}{2}$ inches thick. Slaters' and tilers' laths.

BATTEN-DOOR.—A ledged or barred door.

BATTENING.—Narrow battens fixed to a wall, to which the laths for plastering are nailed.

BATTER.—To incline from the perpendicular. Thus a wall is said to batter when it recedes as it rises.

BATTLEMENT.—A parapet of a building provided with openings or embrasures, or the embrasures themselves. The portions of wall which separate the embrasures are called merlons.

BAY.—A term applied in architecture without much precision.—1. Any opening in a wall left for the insertion of

a a a, Merlons.   b b, Embrasures.

a door or window.—2. Any distinct recess in a building.—3. The quadrangular space between the principal divisions of a groined roof, over which a pair of diagonal ribs extend, and rest on the four angles.—4. The horizontal space between two principals.—5. The division of a building comprised between two buttresses.—6. The part of a window included between two mullions, called also day or light.—7. In a barn, a low enclosed space for depositing straw or hay; or the space between the threshing-floor and the end of the barn.

BAY-WINDOW.—A projecting window, rising from the ground or basement on a semi-octagonal or some other polygonal plan, but generally understood to be straight-sided. When a projecting window is circular in its plan, it is a bow-window; when it is supported on a bracket or corbel, and is circular or polygonal, it is an oriel.

BEAD.—A small round moulding. A series of beads parallel to and in contact with each other is called a reed.—In joinery the bead is of constant occurrence; when it is flush with the face of the work it is called a quirk-bead; when it is raised, a cock-bead.

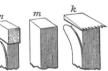

Bay-Window, Glastonbury.

BEAM.—Properly, in carpentry, a piece of timber designed to carry a load applied transversely, as in the case of a beam supporting a floor. In roofs, the tie-beam and collar-beam may or may not be thus transversely loaded.

BEAM-COMPASS.—An instrument used in describing large circles. It consists of a wood or metal beam, having sliding sockets, with steel and pencil or ink points.

BEAM-FILLING.—Filling in between timbers with masonry or brick-work.

BEARER.—In carpentry, a comprehensive term to indicate any supporting member; a beam.

BEARING.—The space between the two fixed extremes of a piece of timber; the unsupported part of a piece of timber; also, the length of the part that rests on the supports.

BEDDING-IN.—The operation of fixing door-frames, window-frames, &c., in walls, and making a tight joint between them and the masonry or brickwork, usually with lime-mortar.

BED-MOULDING.—Properly those members of a cornice which lie below the corona.

BEETLE.—A heavy wooden maul or hammer.

BELL.—The body of a Corinthian or Composite capital, supposing the foliage stripped off.

BELL-GABLE.—In small Gothic churches and chapels, a kind of turret placed on the apex of a gable, usually at the west end, and carrying a bell or bells.

BELL-ROOF.—A roof shaped like a bell, its vertical section being a curve of contrary flexure.

BELVIDERE.—A room placed at the top of a house in order to command a wider prospect.

BENCH.—A strong table on which artisans prepare their work. A cabinet-maker's bench is shown in the figure. The framing is connected by screw-bolts and nuts. The top surface is a thick plank planed very true. It has a trough at a to receive small tools, and a drawer at z. Two side-screws c d, with the chop e, constitute a vice for fixing work. An end-screw g, and sliding piece h, form another vice for thin works which require to be held at right angles to the position of the other chop e; but its chief use is to hold work by the two ends. Work, when laid on the top of the bench, is steadied by the iron bench-hook k, which slides in a mortise in the top, and has teeth at the end, which catch the wood. When work would be injured by the bench-hook, the stop m, sliding stiffly in a square mortise in the bench-top, serves to stay it. There are several square holes along the front of the top, at distances apart from each other equal to the motion of the sliding-piece h, which has a similar hole. In these bench-holes the iron stop n is inserted, and a similar stop is also inserted in the hole in h. Thus, any piece of wood whose length does not exceed the distance between the end-hole of the bench and the stop in h when it is drawn out to the full extent of its range, may be secured. The face of the stop n is slightly roughened. A holdfast o, sliding loosely in a mortise, is used in holding square pieces of work on the bench. It is fixed by driving on the

Iron stop.   Stop.   Bench-hook.

Cabinet-maker's Bench.

top, and released by driving on the back. At *p* is a pin, which is placed in any of the holes shown in the piece in which it is fixed, to support the end of long pieces, which are held by the screws *c d* at their other extremity. Various improvements in the bench-hooks, stops, and holdfasts have been from time to time suggested, such as making them work by screws, but have not obtained general use.—The carpenter's bench is composed of a platform or top, supported on strong framing. It is furnished with a bench-hook at the left-hand end; at which end also the side-board has the screw and screw-cheek, together forming the vice or bench-screw. The side-board and right-hand leg of the bench are pierced with holes, into any one of which a pin is inserted, to hold up the end of any long piece of work clamped in the bench-screw. The length of the bench may be 10 to 12 feet, the breadth 2 feet 6 inches, the height about 2 feet 8 inches. The legs should be 3½ inches square, well braced; front top-board should be 2 inches thick; the further boards may be 1¼ inch.

BENCH-PLANES.—The jack, trying, long, jointer, smoothing, block, and compass planes.

BEVEL.—An instrument for drawing angles. It consists of two limbs jointed together, one called the *stock*, and the other the *blade*, which is movable on a pivot at the joint, so that it may be adjusted to include any angle between it and the stock.

BEVEL-EDGE.—A sloping or inclined edge.

BEVEL-TOOLS.—Tools used in turning hard woods. They are in pairs, and their cutting edges are bevelled off right and left.

BILLET.—An ornament much used in Norman architecture.

Billet-Moulding.

BINDING-JOISTS.—Beams in framed floors which support the bridging-joists above and the ceiling-joists below.

BINDING-RAFTERS.—The same as *Purlins*.

BIRD'S-MOUTH.—An interior angle or notch cut across the grain at the extremity of a piece of timber, for its reception on the edge of another piece.

BIT.—1. The cutting part of a plane. —2. A name common to all those exchangeable boring tools for wood applied by means of the crank-formed handle known as the carpenter's brace.

BLOCK AND PULLEY, Block AND Tackle.—A simple kind of hoisting tackle consisting of a set of pulleys in blocks or sheaves, with a rope or chain passing over the pulleys.

Bits.—a, Shell-Bit; b, Centre-Bit; c, Rose-Bit.

BLOCKINGS or Blocks.—Small pieces of wood glued to the interior angle of two meeting boards to strengthen the joint. Also, small pieces of wood used for wood-block flooring.

BOARD.—A piece of thin timber of considerable length and breadth as compared with its thickness. *Match*-boards are boards with the two edges finished in different ways, but so that the right-hand edge of one fits into the left-hand edge of the other. —*Planed and thicknessed* boards are wrought to a smooth surface and brought to uniform thickness.— *Rough* boards are in the state in which they leave the saw.—*Sound-boarding* is boarding fixed between floor-joists about 2 inches below their upper surface, and covered with some kind of pugging—mortar, slag-wool, &c. —*Weather-boarding* is used on external framing, and is often thicker on the lower edge than the upper, the boards being laid horizontally and overlapping.

BOLECTION-MOULDINGS.—Mouldings in framed work which project beyond the surface of the framing. Called also *balection* and *belection*.

BOLT.—A cylindrical piece of wrought-iron for fastening together the parts of framing or machinery. Usually the bolt has a *head* at one end and a screw thread at the other; the bolt is fixed by means of a *nut* on the screwed end. Also, a round or square bar of metal, sliding in a barrel or through loops into a hole or socket, and used for fastening doors, &c.

BOND-TIMBER.—Timber placed in horizontal tiers at certain intervals in the walls of buildings for attaching battens, laths, and other finishings of wood.

BONING.—The act of judging of a plane surface, or of setting objects in the same plane or line by the eye.

BOSS.—An ornament placed at the intersection of the ribs of groined or cross-vaulted roofs, &c.

Boss, York Cathedral.

BOW-SAW; called also Frame-Saw and Sweep-Saw.—It is used for cutting curves. The frame consists of a central rod or stretcher, to which are mortised two end pieces that have a slight motion of rotation on the stretcher. These end pieces are each adapted at one extremity to receive the saw-blade, and the other ends are connected by a coil or string, in the middle of which is a short lever. On turning round the lever, the string is twisted, and thereby shortened. It thus draws together those ends of the cross pieces to which it is attached, and separates the opposite ends, by which means the saw is stretched.

BOXED SHUTTERS.—Those which fold into a box or case.

BOXINGS of a Window.—The cases into which the shutters of windows are folded.

BRACE.—1. See Angle-Brace.—2. An instrument made of wood or iron, consisting of a cranked shaft, having at its one end a socket, called the *pad*, to receive the bits or boring tools, and at the other a swivelled head or shield, which, when the instrument is used horizontally, is pressed forward by the workman's breast, and when vertically, by his left hand, which is commonly placed against his forehead.

BRACKET.—A small support against a wall for a figure, clock, &c. Brackets, in joinery, are either cut out of deal or framed with three pieces of timber, viz. a vertical piece attached to the wall, a horizontal piece attached to the shelf to be supported, and an angle-brace framed between the horizontal and vertical pieces. Generally, any cantilever or projecting support. The rough pieces of wood fixed to receive the laths for large moulded plaster cornices.

BRACKETED STAIRS.—Stairs with cut strings and small ornamental brackets on the ends of the steps.

BRAD.—A particular kind of nail, used in floors or other work where it is deemed proper to drive nails entirely into the wood. To this end it is made with a small projection on one side at the top, instead of a broad head.

BRAD-AWL.—An awl used to make holes for brads.

BRANDERING. — Covering the under side of joists with battens about 1 inch square in the section, and 12 to 14 inches apart, to nail the laths to, in order to secure a better key for the plaster of a ceiling.

BRANDISHING or Brattishing. — A crest, battlement, or other parapet.

BREAK.—A recess; also, any projection from the general surface of a wall or building.

BREAKING-JOINT.—That disposition of joints by which the occurrence of two contiguous joints in the same straight line is avoided.

BREAST.—A projecting portion of a wall, as a chimney-breast.

BREAST-LINING.—The wood lining between the floor and window-board.

BREEZE-BRICK.—A brick made of coke-breeze and Portland cement, and capable of receiving and holding nails for attaching skirtings and other joinery. It is built into the ordinary brick-work.

BRESSUMMER or Breastsummer.—A beam used in the face or breast of a wall, to support a superincumbent wall.

BRICK, Wood-Brick.—A piece of wood the size of an ordinary brick, but about ⅛ inch thicker, built into a wall to receive the nails for securing door-frames, &c. Wood-bricks work loose in consequence of shrinkage, and fixing-blocks (usually of coke-breeze concrete) are now often used in place of them.

BRICK-NOGGING. — Brickwork carried up and filled in between timber framing.

BRICK-TRIMMER.—A brick arch abutting against the wood trimming joist in front of a fireplace, and used to support the hearth.

BRIDGE-BOARD or Notch-Board. — A board into which the ends of wooden steps are fastened.

BRIDGE-GUTTER.—A gutter formed of boards covered with lead, supported on wooden bearers.

BRIDGING-JOISTS.—The joists to which the flooring boards are nailed.

BRIDGINGS.—Pieces of wood placed between two beams or other timbers, to prevent their approaching each other. More generally termed *straining* or *strutting* pieces.

BROACH.—1. A general name for all tapered boring bits or drills. Those for wood are fluted like the shell-bit, but tapered towards the point. Broaches are also known as *wideners* and *rimers*.—2. An old English term for a spire; in some other parts of the country it is used to denote a spire springing from the tower without any intermediate parapet.

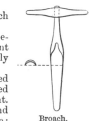

Broach.

BROAD.—An edge tool for turning soft wood. The edge of the broad is at right angles to the handle, and the blade is either

square or triangular. The triangular broad is used principally for turning large pieces the plank way of the grain.

BUHLWORK.—In cabinet-making, furniture inlaid with coloured woods, imitation pearls, &c.

BUILT BEAM.—One composed of two or more timbers placed side by side and fixed together with bolts, keys, &c.

BULLER-NAILS.—Round-headed nails with short shanks, turned and lackered; used chiefly for the hangings of rooms.

BULL-NOSE.—A salient angle with the corner rounded off.

BULL'S-EYE.—A small circular or elliptical window.

BUNDLE.—A number of small articles tied together. Slaters' and plasterers' laths are sold by the "bundle".

BUTMENT-CHEEKS.—The parts at the sides of a mortise.

BUTT END OF TIMBER.—That nearest the root of the tree.

BUTT-HINGES.—Those placed on the edges of doors, &c., with their knuckle on the side on which the door opens.

BUTT-JOINT.—See ABUT.

BUTTON.—A small knob fixed to a door, window, &c. A small circular ornament. A small metal fastener turning on a central pivot.

BUTTRESS.—1. A prop.—2. A projection from a wall to impart additional strength and support. See FLYING-BUTTRESS.

BYRE.—A building in which cows are kept; a *mistal*.

BYZANTINE ARCHITECTURE.—A style of architecture developed in the Byzantine Empire about A.D. 300, and which, under various modifications, continued in use till the final conquest of that empire by the Turks in A.D. 1453.

C

CABIN-HOOK.—A hook pivoted at one end and fitting into an eye at the other.

CABLE-MOULDING.—A cylindrical moulding inserted in a flute so as partly to fill it. In mediæval architecture the cable is a moulding of the torus kind, carved in imitation of a rope.

CAISSON.—A sunk panel in a ceiling or soffit. See COFFER.

Cable-Moulding.

CALIBRE, CALIPER COMPASSES. — Compasses with arched legs, to take the dimensions of the exterior diameter of round bodies; compasses with straight legs, with their points retracted, used to measure the interior diameter or bore of a cylinder.

CAMBER.—A curve or arch.—*Cambered beam*, a beam bent or cut in a slight curve.

CAMPANILE.—A clock or bell tower. The term is more especially applied to detached buildings erected for the purpose of containing bells.

CAMP-CEILING.—1. The interior of a truncated pyramid.—2. The ceiling of an attic room where all the sides are equally inclined from the wall to meet the horizontal part in the centre.

Calibre Compasses.

CANAL.—The same as *flute*.—*Canal of the larmier*, the hollow made in the soffit or under side of a cornice.—*Canal of a volute*, a channel in the face of the circumvolutions of the Ionic capital, enclosed by a list or fillet.

CANKER.—A disease in trees.

CANOPY.—1. A decoration serving as a hood or cover suspended over an altar, throne, chair of state, pulpit, and the like. —2. The ornamental projecting head of a niche or tabernacle.— 3. The label moulding or drip-stone surrounding the head of a door or window when ornamented.—4. The metal hood or bonnet over a dog-grate, &c.

CANT.—*v.* To truncate or cut off the external angle formed by the meeting of two planes. Also, to turn over anything on its angle.—*n.* An external or salient angle.

CANTED COLUMN.—A column polygonal in section.

CANTILEVER.—Wooden or other blocks framed into walls, &c., and projecting so as to carry a cornice, balcony, &c.

CANT-MOULDING.—Any moulding with a bevelled face.

CAP.—The congeries of mouldings which forms the head of a pier or pilaster. In joinery, the uppermost of any assemblage of parts.

CAPITAL.—The uppermost part of a column, pillar, or pilaster, usually moulded and often also carved.

CAPPING PIECE.—Any horizontal timber which extends over upright posts, and into which the posts are framed.

CARCASE.—Generally, the frame or main parts of a thing unfinished and unornamented.—*Carcase-flooring*, the frame of timbers which supports the floor-boards above and the ceiling

below.—*Carcase-roofing*, the frame of timber-work which spans the building, and carries the slate boarding and other covering.

CARPENTER'S RULE. — A folding rule of boxwood 2 or 3 feet in length.—*Carpenter's square*, see SQUARE.

CARRIAGE.—The timber frame which supports the steps of a wooden stair.

CARTOUCHE.—1. A roll or scroll.—2. A tablet formed like a sheet of paper, with the edges rolled up, either to receive an inscription or for ornament.—3. A kind of block or modillion used in the cornices of apartments.

CARYATIDES.—Figures of women in long robes, used in the place of columns as supports for an entablature.

CASE-BAGS.—The joists framed between a pair of girders in naked flooring.

CASED SASH-FRAMES. — Sash-frames formed like boxes to receive ropes, weights, and pulleys, and in which the sashes slide freely up and down.

CASEMENT.—1. A compartment between the mullions of a window, but more generally a glazed sash or frame hinged to open like a door.—2. A hollow moulding equal to one-sixth or one-fourth of a circle.

Caryatid.

CASTING, WARPING, or BUCKLING.—The bending of the surfaces of a piece of timber from their original state, caused either by the weight of the material or by unequal temperature, unequal moisture, or the want of uniformity of texture.

CATHETUS.—A perpendicular line supposed to pass through the middle of a cylindrical body; the axis of a cylinder; the centre of the Ionic volute.

CAVETTO.—A hollow member or concave moulding.

CEILING.—The plaster or other covering which forms the roof of a room.—*Ceiling-joists*, joists to which the ceiling of a room is attached.

CENTRE or CENTRING.—The mould or timber frame on which any arched or vaulted work is constructed.

CENTRE-BIT.—A tool for boring large circular holes.

CESS-PIT.—A small square box formed in a lead gutter, &c., to which the rain-water drains, and from which the rain-water pipe is taken.

CHAIN-TIMBERS.—Bond timbers of a larger size than usual, introduced to tie and strengthen a wall.

CHAIR-RAIL.—A wood rail attached to a wall to prevent injury to the plaster from the backs of chairs.

CHAMFER.—To cut in a slope.

CHANTLATE. A piece of wood fastened near the end of a rafter, or at the foot of tile-hung framing, and projecting beyond the wall, to support two or three rows of slates or tiles, so placed as to prevent the rain-water trickling down the wall.

CHAPLET.—A small cylindrical moulding carved into beads and the like.

CHAPS.—A defect in trees.

CHASE-MORTISE or PULLEY-MORTISE.—A manner of mortising transverse pieces into parallel timbers already fixed. One end of the transverse piece is mortised into one of the parallel pieces, and a long mortise being cut in the other parallel piece, the other end of the transverse piece is let into it, by making it radiate on its already mortised end.

CHECK.—A small hollow or groove in the rebate of a door-frame or casement-frame to check the passage of rain.

CHEEKS.—The sides of a mortise, dormer, &c.

CHEVAL-DE-FRISE.—A piece of timber pointed with iron and transversed with spikes.

CHEVRON. — A carved decoration, consisting of mouldings ranging in zigzag lines, common in the Norman style of architecture. It is called also *zigzag* and *dancette*.

CHISEL.—A cutting tool. —*Carpenters' chisels*: 1. The socket or heading chisel, employed in cutting mortises.

Chevron Moulding.

Its blade is from 1¼ inch to 1½ inch wide, and the top of its stem is formed into a socket to receive a wooden handle. 2. The mortise-chisel, which has a button on the top of its stem, with a tang for insertion into a wooden handle. 3. The ripping chisel, which is generally an old socket-chisel.—*Joiners' chisels*: 1. The mortise chisel, the same as that of the carpenter's, and of various sizes. 2. The firmer. 3. The paring-chisel. 4. The drawing-knife, which is an oblique-ended chisel.

CHORDS OF A TIMBER BRIDGE.—The horizontal longitudinal main timbers of the framing.

CINCTURE.—A ring or list at the top and bottom of a column,

separating the shaft at the one end from the base, and at the other from the capital.

CINQUE-FOIL.—An ornament in Gothic architecture, consisting of five cusped divisions. See FOLIATIONS.

CIRCLE, WORK ON ———.—Framing, the plan of which is a segment of a circle, as a window in a segmental bay.—*Circle-on-circle* work is framing, the plan and elevation of which are segments of a circle, as a semicircular or circular window in a segmental bay.

CIRCULAR SAW.—A saw with circular blade mounted on a spindle like a wheel, with teeth on its periphery.

CISTERN.—A receptacle for storing water. Small cisterns of wood, such as those used for flushing water-closets, are dovetailed at the angles. In large cisterns the sides are housed and bolted together.

CLAMP.—An instrument made of wood or metal, with a screw at one end, used to hold pieces of timber together until the glue hardens, and in laying floor-boards, &c.; also, a piece of wood fixed to another in such a manner that the fibres cross, and thus prevent casting or warping.

CLAP-BOARD.—1. A weather-board on the side of a house, lapping the one beneath it, clinker fashion.—2. A roofing board, larger than a shingle.

CLASP-NAILS.—Nails with heads flattened, so as to clasp the wood.

CLAW-HAMMER.—A hammer having a bifurcated bent peen, suitable for catching below the head of a nail in order to draw it out.

CLEAN.—Free from knots and other defects.

CLEAR.—Free from interruption.—*In the clear*, the net distance between any two bodies.—*Clear-story*, the part of the nave, choir, and transepts of a church above the roofs of the aisles.

CLEARCOLE or CLAIRCOL.—A composition of size and white-lead.

CLEAT.—A strip of wood secured to another to strengthen it, as a batten placed transversely on the back of several boards which are jointed or matched together.

CLEAT-HOOK.—A kind of double hook to which a cord can be secured by winding.

CLENCH, CLINCH.—When two pieces are secured together by nails of greater length than the thickness of the pieces, and the projecting points are turned back and hammered into the wood, the nails are said to be *clenched*, and the pieces to be *clench-nailed*.

CLOSE-STRING.—A stair-string into which the ends of the treads and risers are housed.

CLOUT-NAIL.—A flat-headed nail with which sheet-iron or felt is usually fastened to wood.

CLUSTERED COLUMN.—A pier formed of a congeries of columns clustered together, either attached or detached.

COACH-SCREW.—A large screw, usually with a square head, like that of a bolt.

COAK, COG.—1. A projection from the general face of a scarfed timber of the nature of a tenon, and occupying a recess or mortise in the counterpart face of the other timber; a tabling.—2. A joggle or dowel by which pieces are united to prevent slipping past each other.

COCK-BEAD.—A bead which projects from the surface of the timber on both sides of it.

COFFER.—A panel deeply recessed in any soffit or ceiling.

COGGING, COCKING.—A mode of notching timber; called also *caulking*.

Cogging.

COIN, COIGNE.—The corner of a building. See QUOIN.

COLLAR OF A SHAFT.—The annulet.

COLLAR-BEAM.—A beam extending between the two opposite rafters of a framed principal above the tie-beam, or between two common rafters.

COLONNADE.—A range of columns.

COLUMN.—A long solid body called a shaft, usually set on a congeries of mouldings, which forms its base, and surmounted by a spreading mass, which forms its capital. Columns are distinguished by the styles of architecture to which they belong, and, in classic architecture, by the name of the order; and again by some peculiarity of position, construction, form, or ornament—as attached, twisted, cabled, indented, rusticated columns.

COMMON JOISTS.—Those in naked flooring to which the boards are attached, called also *bridging-joists*.

COMMON RAFTERS.—Those to which the slate-boarding or lathing is attached.

COMMON ROOFING.—Roofing consisting of common rafters only, without principals.

COMPASS-HEADED ARCH.—A semicircular arch.

COMPASS-PLANE.—A plane with a round sole.

COMPASS-ROOF.—One in which the tie from the foot of one rafter is attached to the opposite rafter at a considerable height above its foot.

COMPASS-SAW.—A saw with a narrow blade adapted to run in a circle of moderate radius. By a rotation of the hand it is constantly swerved, and its kerf allows it some play, so that it cuts in a curve. It is usually thick enough on the cutting edge to run without any set. The blade is usually an inch wide at the handle, tapers to a quarter of an inch at the point, and has five teeth to the inch. It is also called *fret-saw*.

COMPASS-WINDOW.—A bay-window on a circular plan.

COMPOSITE ORDER.—The last of the five Roman orders. Its capital has a vase like the Corinthian, surrounded by two rows of acanthus leaves; the top of its vase is surmounted by a fillet, astragal, and ovolo; over this the volutes roll angularly, till they meet the tops of the upper row of leaves, on which they seem to rest. On the top of each volute is an acanthus leaf, curling upwards so as to sustain the horn of the abacus.

Composite Capital.

CONCENTRIC.—Having a common centre, as concentric circles, ellipses, spheres.

CONE.—A right cone is a solid with a circular base, and tapering upwards uniformly to a point known as the vertex. The five sections of a cone are the circle, triangle, ellipse, parabola, and hyperbola.

CONGÉ.—The cavetto which unites the base and capital of a column to its shaft.

CONSOLE.—A projecting ornament used as a bracket. It has for its outline generally a curve of contrary flexure.

CORBEL.—*n.* A structure of stone, brick, wood, or iron, incorporated with a wall, and projecting from its vertical face, to support some superincumbent member.—*v.* To dilate by expanding every member of a series beyond the one under it.

Cornice supported by Consoles.

CORBEL-STEPS.—Steps into which the sides of gables from the eaves to the apex are broken; sometimes called *corbie-steps*.

CORBEL-TABLE.—A term in mediæval architecture, applied to a projecting course and the row of corbels which support it.

CORINTHIAN ORDER.—The only Grecian example of this order remaining is in the choragic monument of Lysicrates at Athens. (See SECTION I.) The principal feature is the foliated capital. The Corinthian order was freely used by the Romans.

Corbel-Table.

Among the examples is the temple of Jupiter Stator at Rome, in which the column is 10 diameters high, the base occupying of this $\frac{1}{2}$ diameter, and the capital $1\frac{1}{6}$ diameter; and the entablature is more than $2\frac{1}{2}$ diameters in height, the architrave and frieze having about $1\frac{1}{2}$ diameter between them.

CORNICE.—1. The highest part of an entablature resting on the frieze.—2. Any congeries of mouldings which crowns or finishes a composition externally or internally.

CORONA.—A member of a cornice situated between the bed-moulding and the cymatium. It consists of a broad vertical face, usually of considerable projection. Its soffit is generally recessed upwards to facilitate the fall of rain from its face. This among workmen is called the *drip*, and by the French *larmier*.

Corinthian Capital—Roman.

CORPSE-GATE.—See LICH-GATE.

CORPSING.—A shallow mortise sunk in the face of a piece of stuff.

COUNTER-LATH.—A lath, in tiling, placed between every two gauged ones so as to make equal spaces. In plastering, a lath nailed to a beam or other timber before the plastering laths are nailed on.

COUNTERLIGHT.—A glazed sash or top-light placed horizontally in a ceiling, under a sky-light in the roof.

COUNTERSINK, *v.*—To form a cavity in timber or other material for the reception of something, such as the head of a bolt.

COUPLED COLUMNS.—Columns disposed in pairs. The two columns of a pair are usually half a diameter apart.

COUPLES, COUPLE CLOSE.—A pair of opposite rafters in a roof nailed together at the top, where they meet, and connected by a tie at the bottom, or by a collar-beam higher up.

COVE.—Any kind of concave moulding. The concavity of a vault; commonly applied to the curve which is sometimes used to connect the ceiling with the walls of a room.

COVE BRACKETING.—The wooden skeleton mould or bracketing of a coved ceiling.

CRAB.—A winch or small crane.

CRADLE.—A name given to a centring of ribs latticed with spars, used in building culverts.

CRADLE-VAULT.—An improper term for a cylindrical vault.

CRADLING.—Timber framing for sustaining the lath and plastering of a vaulted ceiling or around an iron beam, &c. Also, the framework to which the wooden entablature of a shop front is attached.

CRAMP.—A piece of iron bent at the ends, serving to hold together pieces of timber, stone, &c.; a small apparatus used to press boards tightly together before nailing.

CRAMPOONS.—An apparatus used in the raising of timber or stones, consisting of two hooked pieces of iron hinged together, somewhat like double calipers.

CRENELLATED MOULDINGS.—Mouldings embattled, notched, or indented, used in the Norman style.

CRESTS.—Carved work on the top of a building. The ridges of roofs, the copes of battlements, and the tops of gables were also called *crests*.

Crenellated or Embattled Moulding.

CRIPLINGS.—Spars set up as shores against the sides of buildings.

CROCKET.—In Gothic architecture, an ornament placed at the angles of pinnacles, gables, canopies, and other members. In its usual form the crocket is a foliated band, covering the angle of the member to which it is applied, swelling out at regular intervals into knobs with considerable projection.

CROSS - GARNETS. — Hinges having a long strap fixed to the door or closure, and a shorter right-angled strap fixed to the frame. Termed in Scotland *cross-tailed hinges*.

Crockets.

CROSS-GRAINED STUFF.—Timber having the grain or fibre not corresponding to the direction of its length, but crossing it, or irregular. Where a branch has shot from the trunk of a tree, the timber of the latter is curled in the grain.

CROSS-SPRINGER.—The diagonal rib of a vault.

CROSS-VAULTING.—That which is formed by the intersection of two or more simple vaults.

CROWN OF AN ARCH.—Its vertex or highest point.

CROWN-POST.—The same as *king-post*.

CUPOLA.—A spherical vault at the top of an edifice; a dome, or the round part of a dome. The Italian word signifies a hemispherical roof, covering a circular building like the Pantheon at Rome or the Round Temple at Tivoli.

Cupola, Radclyffe Library, Oxford.

CURB.—A frame round the mouth of an opening; a curb-plate.—*Curb-plate*: 1. The wall-plate of a circular or elliptical domical roof, or of a sky-light or bay-window. 2. The plate which receives the upper rafters of a curb or Mansard roof. 3. The circular frame of a well.—*Curb-rafters*, the upper rafters of the curb or Mansard roof.—*Curb-roof*, the same as a *Mansard roof*.

CURLING-STUFF.—Timbers in which the fibres wind or curl where boughs have shot out from the trunk of the tree.

CURTAIL-STEP.—The first step of a stair when its outer end is finished in the form of a scroll.

CUSHION-CAPITAL.—A capital having a resemblance to a cushion pressed by a weight; most prevalent in the Norman style. It consists of a cube rounded off at its lower angles.

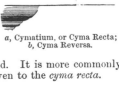

Norman Cushion-Capital.

CUSPS.—The points or small projecting arcs terminating the internal curves of the trefoiled, cinquefoiled, &c., heads of windows and panels in Gothic architecture. They are also called *featherings*.

CUT AND MITRED STRING. A stair-string with notches cut in the upper edge to receive the treads and risers, and mitred to the risers.

CUT - BRACKETS. — Those moulded on the edge.

CUT - ROOF. — A truncated roof.

CYLINDER.—A round solid of uniform diameter, of which the bases are equal and parallel circles. In hand-railing the term has a special meaning.

CYLINDRICAL VAULT. —A vault without either groins or ribs. Its vertical section is a semicircle, or some lesser arc. It is called also *cradle-vault*, *wagon-vault*, and *barrel-vault*.

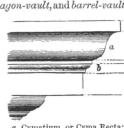

Cusps.

1. Monument of Edward III, Westminster Abbey (brass). 2. Henry VII's Chapel. 3. Monument of Sir James Douglas, Douglas Church. 4. Beauchamp Chapel, Warwick.

CYLINDRIC GROINS.—Those produced by the intersection of cylindric vaults of equal span and height.

CYLINDROID.—A solid, which differs from a cylinder in having ellipses instead of circles for its ends or bases.

CYMA RECTA or CYMATIUM.— A moulding formed of a curve of contrary flexure, concave at the top and convex at the bottom.

a, Cymatium, or Cyma Recta; *b*, Cyma Reversa.

CYMA REVERSA.—A moulding formed of the above curve reversed. It is more commonly called an *ogee*, but this name is also given to the *cyma recta*.

D

DADO.—That part of a pedestal included between the base and the cornice; the die.—In apartments, the dado is that part of the finishing between the base and the surbase.

DAIS.—1. A platform or raised floor at the upper end of an ancient dining-hall, where the high table stood.—2. A seat with a high wainscot back, and sometimes with a canopy, for those who sat at the high table.—3. The high table itself.

DANCE.—In a stair of mixed fliers and winders, the distribution of the inequality of width of the inner end of the latter among them and a number of the fliers, is called *making the steps dance*.

DAY.—One of the divisions of a window contained between two mullions. In this sense, the same as *bay*.

DEAD-SHORE.—See SHORE.

DEAFENING.—Anything used to prevent the passage of sound in floors or partitions. The term used in Scotland as synonymous with *pugging*.

DEAL.—The usual thickness of deals is 3 inches and width 9 inches. *Red* or *yellow deal* is the wood of the Scotch fir (*Pinus silvestris*); *white deal* that of the spruce (*Picea excelsa*).

Pedestal.

b, Dado or Die. *a*, Surbase. *c*, Base.

DECORATED STYLE.—The second of the Pointed or Gothic styles of architecture used in this country. It was developed

York Minster, West Front.

from the Early English at the end of the thirteenth century. The most characteristic feature is the tracery, which is always either of geometrical figures, circles, quatrefoils, &c. (in the earlier examples), or flowing, wavy lines (in later specimens). The west front of York Minster is a good example, but the upper part of the towers is in the Perpendicular style.

DECORATION. — Anything which adorns and enriches an edifice, as vases, statues, paintings, festoons, &c.

DEFLECTION.—The bending or *sagging* of a beam.

DEMI-RILIEVO. — Sculpture in relief, in which one-half of the figure projects from the plane of the stone or other material from which it is carved. It is also called *mezzo-rilievo.*

DENTILLED, DENTICULATED.—Having dentils.

DENTILS.—Ornaments in the form of little cubes or teeth, used in the bed-moulding of Ionic, Corinthian, and Composite cornices.

DERRICK.—A kind of crane much used in building for lifting heavy weights, and swinging them into position.

DIAGLYPHIC.—A term applied to sculpture, engraving, &c., in which the objects are sunk below the general surface.

DIAGONAL.—A right line drawn from angle to angle of a four-sided figure.

DIAGONAL SCALE.—A special scale for measuring with great accuracy.

DIAMETER.—1. A right line passing through the centre of a circle or other curvilinear figure, terminated by the circumference, and dividing the figure into two equal parts.—2. In architecture, the measure across the lower part of the shaft of a column, which, being divided into 60 parts, forms a scale by which all the parts of the order are measured. The 60th part of the diameter is called a minute, and 30 minutes make a module.—3. A right line passing through the centre of a piece

Decorated Capital, Selby, Yorkshire.

Dentils.

Diamond Fret.

of timber, a rock, or other object, from one side to the other; as, the *diameter* of a tree or of a stone.

DIAMOND FRET.—A decorated moulding consisting of fillets intersecting each other, so as to form diamonds or rhombuses; used in Norman architecture.

DIAPER.—Ornament of sculpture in low relief, sunk below the general surface, or of painting or gilding, used to decorate a panel or other flat recessed surface.

Diaper, Westminster Abbey.

DICOTYLEDONOUS or EXOGENOUS TREES.—See EXOGEN.

DIE.—The part of a pedestal between its base and cornice; a dado.

DIRECTORS.—Triangular compasses.

DIRECTRIX.—A line perpendicular to the axis of a conic section, to which the distance of any point in the curve is to the distance of the same point from the focus in a constant ratio; also the name given to any line, whether straight or not, that is required for the description of a curve. Thus A B is the directrix of the parabola V E D, of which F is the focus.

Directrix.

DISCHARGING ARCH.—An arch formed in the substance of a wall, to relieve the part below it of the superincumbent weight. Such arches are commonly used over lintels and flat-headed openings, and are known as *relieving-arches.*

DISHED.—Formed in a concave.—*To dish out,* to form coves by wooden ribs.

DODECAHEDRON.—A solid figure, consisting of twelve equal sides.

DOG-LEGGED STAIRS.—Stairs in which the outer string of the upper flight is vertically over that of the lower flight.

DOG'S-TOOTH MOULDING. — An ornamental member, very characteristic of Early English architecture. See TOOTH-ORNAMENT.

DOME.—The hemispherical cover of a building; a cupola.

DOOR-CASE, DOOR-FRAME.—The frame around a door.—*Door-nail,* the nail or knob in ancient doors on which the knocker struck.—*Door-post,* the post of a door.—*Door-stops,* pieces of wood against which the door shuts in its frame.

DORIC ORDER.—The earliest and simplest of the three Greek orders. It may be divided into three parts, stylobate, column, and entablature. The Roman-Doric may be regarded as a deteriorated imitation of the Grecian-Doric.

DORMANT, DORMANT-TREE, DORMAR.—A summer, sommer, somnier, or sleeper; a beam.

DORMER-WINDOW.—A window in the sloping side of a roof, with its casement set vertically. When the window lies in the plane of the roof, it is called a *sky-light.*

DORSE.—A canopy.

DOUBLE-HUNG SASH.—A sash-window in which both the upper and the lower sashes are suspended with cords or lines passing over pulleys and attached to balance-weights; in a *single-hung* sash, only one of the sashes is so suspended, the other being fixed.

DOUBLE-JOISTED FLOOR.—A floor with binding and bridging joists.

DOUBLE-MARGINED DOORS.—Doors in which the muntin is double the width of the stiles and divided into two parts by a vertical bead or mould.

DOUBLE SCALES.—The sector, &c.

DOVE-TAIL.—The manner of fastening boards and timbers together, by letting one piece into the other, in the form of a dove's tail spread or a wedge reversed, so that it cannot be drawn out in the direction of its fibres.

DOVE-TAIL MOULDING.—A moulding decorated with running bands in the form of dove-tails, used in Norman architecture. It is sometimes called *triangular fret.*

DOWEL. — To fasten boards together by pins inserted in their edges, or to secure posts to bases, sills, and beams by pins. The pin of wood, iron, copper, slate, &c., used in dowelling.

Dove-tail Moulding.

DRAGON-BEAM, DRAGON-PIECE, DRAGON-TIE.—A beam or piece of timber bisecting the angle formed by the meeting of the wall-plates of two sides of a building, used to receive and support the foot of the hip-rafter.

DRAUGHT.—1. In masonry, a line on the surface of a stone hewn to the breadth of the chisel.—2. In carpentry and joinery,

when a tenon is to be secured in a mortise by a pin passed through both pieces, and the hole in the tenon is made nearer the shoulder than to the cheeks of the mortise, the insertion of the pin *draws* the shoulder of the tenon close to the cheeks of the mortise, and it is said to have a *draught*.

DRAW-BORE.—A hole pierced through the tenon, nearer to the shoulder than the holes through the cheeks from the abutment in which the shoulder is to come into contact.

DRAW-BORE PIN.—A joiner's tool, consisting of a solid piece or pin of steel, tapered from the handle, used to enlarge the pin-holes which are to secure a mortise and tenon, and to bring the shoulder of the rail close home to the abutment on the edge of the style. When this is effected, the draw-bore pin is removed, and the hole filled up with a wooden peg.

DRAWING-KNIFE.—An edge-tool used to make an incision into the surface of wood, along the path the saw is to follow. It prevents the teeth of the saw tearing the surface of the timber.

DRESSED.—Planed or wrought.

DRESSINGS.—All mouldings which are applied as ornaments, and project beyond the naked face of the work.

DRIFT.—A piece of iron or steel-rod used in driving back a key of a wheel, or the like, out of its place, when it cannot be struck directly with the hammer.

DRIFT-BOLT.—A long pointed spike, usually with a head like a bolt, and secured by driving into timbers.

DRILL, *v.*—To bore a hole in stone or metal by means of a drill or boring-machine.

DRIP.—The edge of a roof; the eaves; the corona of a cornice; any ledge or projection to prevent rain running down a wall; a change of level in a lead gutter or flat.

DRIPSTONE.—The label moulding in Gothic architecture, which serves as a canopy for an opening, and to throw off the rain. It is also called *weather-moulding* and *water-table*.

Dripstone, Westminster Abbey.

DROPS.—See GUTTÆ.

DRUM.—1. The stylobate or vertical part under a cupola or dome.—2. The solid part of the Corinthian and Composite capitals; called also *bell*, *vase*, *basket*.

DRUXY.—An epithet applied to timber with decayed spots or streaks of a whitish colour in it.

DURAMEN.—The heart-wood of a tree.

DWANGS.—The Scotch term for struts inserted between the joists of a floor or the quarterings of a partition, to stiffen them.

DWARF-WALLS.—Walls of less height than a story of a building. The term is generally applied to the low *sleeper* walls built under the ground-floor, to support the sleeper-joists.

E

EARLY ENGLISH ARCHITECTURE.—The first of the

North-west Transept, Beverley Minster.

Pointed or Gothic styles of architecture that prevailed in this country. It succeeded the Norman towards the end of the twelfth century, and gradually merged into the Decorated at the end of the thirteenth.

EASE, *v.*—To take a thin shaving off any member, such as the edge of a door, for the purpose of making it fit easily.

EAVES.—That part of a roof which projects beyond the face of a wall.—*Eaves-board*; called also *Eaves-catch* and *Eaves-lath*.—An arris-fillet nailed across the rafters at the eaves of a roof, to raise the slates a little.—*Eaves-gutter.*—A gutter or trough attached to the eaves.

Early English Capital, Salisbury Cathedral.

ECHINUS.—An ornament in the form of an egg, peculiar to the ovolo or quarter-round moulding; whence this moulding is sometimes called *echinus*.

EGG-AND-ANCHOR, EGG-AND-DART, EGG-AND-TONGUE.—See ECHINUS.

ELBOW-LINING.—The lining of the elbows or reveals of a window-recess.

Echinus.

ELBOWS.—The upright sides which flank any panelled work, as in windows below the shutters.

ELEVATION.—A geometrical delineation of any object according to its vertical and horizontal dimensions, without regard to its thickness or projections.

ELIZABETHAN ARCHITECTURE.—A name given to the architecture which prevailed in the reigns of Elizabeth and James I., when Gothic and Italian forms were combined. Its chief characteristics are windows of great size, divided into a number of *lights* by mullions and transomes, combined with a profuse use of ornamental strap-work in the parapets, &c.

ELLIPSE.—A section of a cone by an inclined plane passing through the sides of the cone.

EMBATTLEMENT or BATTLEMENT.—An indented parapet, belonging originally to military works, the indents, crenelles, or embrasures being used for the discharge of missiles. It was afterwards adopted as a decoration in mediæval architecture.

Elizabethan Window, Rushton Hall, cir. 1590.

EMBOSS, *v.*—To form bosses or protuberances; to cut or form with prominent figures.

EMBRASURE.—An opening in a wall, splaying or spreading inwards. The term is usually applied to the indent or crenelle of an embattled parapet.

ENDECAGON.—A figure of eleven sides and eleven angles.

ENDOGENS.—Plants whose stems are increased by the development of woody matter towards the centre, instead of at the circumference, as in exogens. To this class belong palms, grasses, rushes, &c. Stems of this sort have no distinct concentric layers or medullary rays.

ENGAGED COLUMN.—A column attached to a wall, so that part of it is concealed. Engaged columns have seldom less than a quarter, or more than a half of their diameter in the solid of the wall.

ENNEAGON.—A polygon with nine sides or nine angles.

ENRICH.—To adorn with carving or sculpture.

ENTABLATURE.—That part of an order which lies upon the abaci of the columns. It consists of three principal divisions—the architrave, frieze, and cornice.

Engaged Column.

ENTASIS.—A swelling; the curved line in which the shaft of a column diminishes; the swelling in the middle of a baluster.

ENTRESOL.—A low story between two other stories.

EQUILATERAL ARCH.—A pointed arch in which the radius of each curve is equal to the span.

EQUILATERAL TRIANGLE.—A triangle with three equal sides and three equal angles.

ESCAPE.—That part of a column where it springs out of the base; the apophyge; the congé.

ESCUTCHEON.—1. A shield for armorial bearings.—2. A plate for protecting the key-hole of a door, or to which the handle is attached.

ESPAGNOLETTE.—A kind of bolt used for folding casements; it consists of an upper and lower rod or bolt, attached to a handle in such a manner that the turning of the handle shoots the two bolts simultaneously up and down.

EXFOLIATION.—The shedding of bark; a disease in trees.

EXOGEN.—A plant whose stem increases by development of woody matter towards the outside. To this class belong all our timber trees.

EXPANDING CENTRE-BIT.—A hand-instrument, chiefly used for cutting out discs of leather and other thin material, and for making the margins of circular recesses. It consists of a central stem *a*, and point *b*, mounted on a transverse bar *c*, which carries a cutter *d* at one end, and is adjustable for radius. The arm *c* being carried round the fixed points *a* and *b*, the cutter *d* describes a circle of which the radius is the distance *b d*.

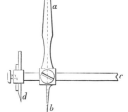

Expanding Centre-bit.

EXTRADOS.—The exterior curve of an arch.

EYE.—A general term applied to the centre of anything, as the eye of a volute or of a dome.

F

FAÇADE.—The face or front of an edifice.

FACE-MOULD.—One of the patterns for marking the board or plank out of which the handrails for stairs and other works are to be cut.

FACETS, FACETTES.—Small projections between the flutings of columns.

FACING.—1. The thin covering of polished stone, or of plaster or cement, on a rough stone or brick wall.—2. The woodwork which is put as a border round apertures, either for ornament or to cover and protect the junction between the frames of the apertures and the plaster.—3. Sometimes in joinery used synonymously with *lining* or *casing*.

FACTOR OF SAFETY.—A number by which the ultimate resistance or breaking weight of any structure, as ascertained by calculation, is divided in order to obtain the *safe* or *working* load.

FALDSTOOL.—A kind of stool placed at the south side of the altar, at which the Kings of England kneel at their coronation.—2. A small desk, at which the Litany is enjoined to be sung or said; sometimes called a *Litany-stool*.—3. The chair of a bishop, enclosed by the railing of the altar.—4. An arm-chair; a folding-chair.

FALLING MOULDS.—The two moulds which, in forming a handrail, are applied, the one to its convex and the other to its concave vertical side, in order to form the back and under-surface, and finish the squaring.

FALLING-STILE.—That stile of a gate or door in which the lock, latch, or other fastening is placed.

Fan-tracery, North Aisle, St. George's Chapel, Windsor.

FALSE ROOF.—The open space between the ceiling of an upper apartment and the rafters of the roof.

FAN-LIGHT.—Properly a semicircular window over the opening of a door, with radiating bars in the form of an open fan, but now used for any window over a door.

FAN-TRACERY VAULTING.—A kind of vaulting in the Perpendicular style, in which the ribs have the same curve, and diverge equally in every direction from the springing of the vault.

FASCIA.—1. A band or fillet.—2. Any flat member with a little projection, as the band of an architrave.—3. A board nailed to the feet of rafters to form a back-plate for the eaves-trough.

FEATHER-EDGED BOARDS.—Boards thinner on one edge than on the other.

FEATHERINGS or FOLIATIONS.—The cusps or arcs of circles with which the divisions of a Gothic window, &c., are ornamented.

FELLOE.—The outer rim of the frame of a centre or mould under the lagging or covering-boards.

FELT.—A fibrous material, usually waterproofed with bitumen, &c., and used for damp courses, and as a covering for roof-boarding. The latter kind is often known as *sarking felt*.

FELT GRAIN.—Timber split in a direction crossing the annular layers towards the centre. When split conformably with the layers it is called the *quarter grain*.

FENCE.—A wall, hedge, railing, paling, &c.; a raised piece on a saw-bench serving as a guide for the timber.

FENDER.—A raised curb, as for a hearth.

FENDER-PILES.—Piles driven to protect work, either on land or water, from the concussion of moving bodies.

FENESTRAL.—A small window. Used also to designate the framed blinds of cloth or canvas that supplied the place of glass previous to the introduction of that material.

FESTOON.—A sculptured ornament in imitation of a garland of fruits, leaves, or flowers, suspended between two points, but sometimes composed of an imitation of drapery similarly disposed, or of an assemblage of musical instruments, implements of war or of the chase, and the like.

FILLET.—A small moulding, generally rectangular in section, and having the appearance of a narrow band. In carpentry and joinery, any small scantling less than a batten.

FILLISTER.—A kind of plane used for grooving timber or for forming rebates.

FINGER-PLATE.—A metal or china plate fixed to a door to protect the paint from the marks of dirty fingers.

FINIAL.—The ornamental termination to a pinnacle, gable, &c., consisting usually of a knot or assemblage of foliage. By old writers finial is used to denote not only the leafy termination, but the whole pyramidal mass.

FIR.—A term somewhat loosely used to designate the wood of the Scotch Fir (*Pinus silvestris*), known also as red or yellow deal.

FIRMER.—A paring-chisel.

FIRRINGS.—Pieces of wood nailed to any range of scantlings to bring them to one plane, applied generally to the pieces added to joists which are under the proper level for laying the floor. Called also *furrings*.

Finial.

FISHING, FISHED BEAM.—A built beam, composed of two beams placed end to end, and secured by pieces of wood or iron, covering the joint on opposite sides, and bolted to the two beams.

FLAMBOYANT STYLE OF ARCHITECTURE.—A term applied by French writers to that style of Gothic architecture in France which was coeval with the Perpendicular style in Britain. Its chief characteristic is a wavy, flame-like tracery in the windows, panels, &c.; whence the name.

FLANK.—1. The side of a building.—2. The Scotch term for a valley in a roof.

FLANNING.—The splaying of a door or window-jamb internally.

FLAPS.—Folds or leaves hinged to window-shutters, tables, &c.

FLASHINGS.—In plumbing, pieces of lead, zinc, or other metal, used to protect the joinings where a roof comes in contact with a wall, or where a chimney-shaft or other object comes through a roof. When the flashing is folded down over the upturned edge of the lead of a gutter, or over the upper surface of slates, tiles, &c., it is termed an *apron*.

Flamboyant Window, Church of St. Ouen at Rouen.

FLATTING.—A coat of paint which, from its mixture with turpentine, leaves the work *flat* or without gloss.

FLÈCHE.—A name for a spire when the altitude is great compared with the base.

FLEUR-DE-LIS.—A heraldic figure representing a lily, the distinctive bearing of the Kingdom of France.

FLIERS.—Steps of a stair which are parallel-sided; such as do not wind.

FLIGHT—A series of fliers from one level to another.

FLITCHED BEAM.—A beam composed of two pieces of timber placed side by side, and bolted together with a flat plate of iron or steel between them.

FLOOR.—1. That part of a building or room on which we walk.—2. A platform of boards, planks, &c., laid on timbers.

FLOORING.—The whole structure of the floor-platform of a building, including the supporting timbers.

FLORIATED.—Having florid ornaments; as, the *floriated* capitals of early Gothic pillars.

FLUE.—A passage for smoke leading from the fireplace to the top of the chimney, or into another passage. Also, a pipe or tube for conveying heat to water, in certain kinds of boilers. Also, a passage for conveying fresh or vitiated air.

FLUING.—Expanding or splaying, as the jambs of a window.

FLUSH.—A term applied to surfaces which are in the same plane.—*Flush panel*, a panel flush with the framing on one or both sides.

FLUTINGS or FLUTES.—The hollows or channels cut perpendicularly in columns, &c. When the flutes are partially filled by a smaller round moulding they are said to be *cabled*.

FLYING-BUTTRESS.—A buttress in the form of an arch springing from a solid mass of masonry and abutting against a higher wall, to resist the thrust of an arch or of a roof.

Flying-buttress, Beverley Minster.

FOCUS.—A point in which any number of rays of light meet, after being reflected or refracted; as, the *focus* of a lens. The *foci of an ellipse* are two points toward the ends of the major axis, from which two right lines, drawn to any point in the circumference, shall together be equal to the longer axis.

FOILS.—The small arcs in the tracery of Gothic windows, panels, &c., which are said to be tre-foiled, quatrefoiled, cinquefoiled, and multifoiled, according to the number of arcs they contain.

Trefoil. Quatrefoil. Cinquefoil.

FOLDED FLOORING.—Flooring with square-edged boards, of which two are first secured at a distance apart slightly less than the width of three or four boards; these are then placed together and forced down into the space and nailed.

FONT.—A vessel employed in Protestant churches to hold water for the purpose of baptism, and in Catholic churches used also for holy water.

FOOTING-BEAM.—The tie-beam of a roof.

FOOT-SPACE.—A landing or resting-place at the end of a flight of steps. If its width equals that of two flights rising in opposite directions, it is termed a *half-space*; if its width is equal to that of one flight only, it is a *quarter-space*.

FORE-PLANE.—The first plane used after the saw or axe.

FORMERETS.—The arches which in Gothic ribbed vaulting lie next the wall, and are consequently only half the thickness of the other ribs.

Font, Colebrooke.

FOUR-CENTRED ARCH.—A pointed arch struck from four centres.

FOX-TAIL WEDGING.—A method of fixing a tenon in a mortise, by splitting the end of the tenon with one or more wedges. When the mortise is not cut through, the wedges are inserted in the end of the tenon, and the tenon entered into the mortise, and then driven home. The bottom of the mortise forces the wedges farther into the tenon, which is thus made to expand, and press firmly against the sides of the mortise.

VOL. II.

FRAME, FRAMING.—A term applied to any assemblage of pieces of timber firmly connected together. Also the rough timbers of a house, including beams, flooring, roofing, and partitions.

FRAMED or BOUND DOORS.—Doors with a surrounding frame, and flush-boarded on one side.

FRAMED-FLOOR.—One with girders, binding-joists, and bridging-joists.

FRENCH DOORS or WINDOWS.—Casement windows adapted for use as doors.

FRET.—A kind of knot or meander, used as an ornament in architecture. See A LA GRECQUE and MEANDER.

FRIEZE.—That part of an entablature which is between the architrave and cornice. It is often enriched with sculptures.

FRIEZE-PANEL.—The uppermost panel of a door or other framing, in which there are three or more tiers of panels.

FRIEZE-RAIL.—The rail next below a frieze-panel.

FRUSTUM.—The part of a solid next the base, left by cutting off the top or segment by a plane parallel to the base.

G

GABLE.—The triangular end of a house or other building, from the cornice or eaves to the top.—*Gable-roof*, a roof open to the sloping rafters or spars, finishing against gable-walls.

GABLET.—A small gable or gable-shaped decoration, frequently introduced on buttresses, screens, &c.

GAIN.—1. A bevelling shoulder.—2. A lapping of timbers.—3. The cut that is made to receive a timber.

GALILEE.—A small gallery or balcony, open to the nave of a conventual church; also a porch or portico annexed to a church, and used for various purposes.

GALLERY.—1. An apartment of much greater length than breadth, serving as a passage of communication for the different rooms of a building, or for the reception of pictures, statues, &c., its use being denoted by a qualifying word, as picture-gallery, sculpture-gallery, &c.—2. A platform projecting from the walls inside a building, and supported on piers, pillars, brackets, or consoles, as the gallery of a church or of a theatre.—3. A long portico, with columns on one side.

GANTRY.—A structure composed of two parallel timber frames, which serve as a support for a travelling crane. The term is also applied to a somewhat similar structure covered with joists and boards and used as a scaffolding.

GARGOYLE or GURGOYLE.—A spout in the cornice or parapet of a building, for throwing the roof-water beyond the walls.

GARNET-HINGE.—A hinge resembling the letter T laid horizontally, thus ⊢; called in Scotland, a *cross-tailed hinge*.

GARRET.—1. That part of a house which is on the upper floor, immediately under the roof.—2. Rotten wood.

GATHERING.—The lower part of the funnel of a chimney.

GAUGE.—The length of the exposed part of a slate or tile when laid; also the measure to which any substance is confined. An instrument used by joiners for making a line on a board parallel to one edge of the board.

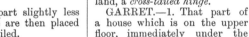

Gargoyle, Stony-Stratford.

GAUGED-PILES.—Large piles placed at regular distances apart, and connected by horizontal beams, called *runners* or *wale-pieces*, fitted to each side of them by notching, and firmly bolted. A gauge or guide is thus formed for the *sheeting* or *filling* piles, which are driven between the wale-pieces. Gauged-piles are called also *standard piles*.

GEOMETRICAL STAIRS.—A term sometimes used to denote stairs which wind around a well or open space.

GIBS.—Pieces of iron employed to clasp together such pieces of a framing as are to be keyed, previous to inserting the keys.

GIRDER.—A main beam to support the joists of a floor. Any large beam of wood, iron, or steel.

GIRTH.—The transverse measurement of the superficies of a cylinder or other body; the girth of a tree is its circumference, and the girth of a series of mouldings is the transverse measurement of the exposed surface.

GOING OF A STEP.—The horizontal distance from the face of one riser to the face of the next. The going of a flight of stairs is the horizontal distance between the first and last risers.

GORE.—A wedge-shaped or triangular piece.

GOTHIC.—In architecture, a term strictly applicable only to the styles which are distinguished by the pointed arch.

GRADE, *n.*—A step or degree.—*v.* To reduce to a certain degree of ascent or descent, as a road or way.

GRADIENT.—The degree of slope of a road, drain, &c.

GRAINING.—Painting in imitation of the grain of wood.

GRATING.—Any perforated plate of wood or metal for the passage of air, water, &c. See GRILLAGE.

GRECIAN ARCHITECTURE. — The architecture which flourished in Greece from about 500 years before the Christian era, or perhaps a little earlier, until the Roman conquest.

GRILLAGE.—A framework composed of beams laid longitudinally, and crossed by similar beams notched upon them, used to sustain walls, and prevent their irregular settling in soils of unequal compressibility. The grillage is firmly bedded and earth or concrete packed in the interstices between the beams, a flooring of thick planks, termed a *platform*, is then laid, and on this the foundation courses of the wall rest.

GROIN.—The line made by the intersection of simple vaults crossing each other at any angle.

GROINED ROOF or CEILING.—A ceiling formed by intersecting vaults, every two of which form a groin at the intersection, and all the groins meet in a common point, called the *apex* or *summit*. The curved surface between two adjacent groins is termed the *sectroid*. See FAN-TRACERY.

GROOVING AND TONGUING, GROOVING AND FEATHERING, PLOUGHING AND TONGUING.—A mode of joining boards by forming a groove or channel along the edge of one board, and a continuous projection or tongue on the edge of the other board.

GROUND-FLOOR.—That floor of a building which is on a level with or a little above the ground without.—*Ground-joists*, joists which rest on dwarf-walls, prop-stones, or bricks laid on the ground; *sleepers.*—*Ground-mould*, an invert mould, used in tunnelling operations, or any mould by which the surface of ground is formed.—*Ground-plan*, the plan of the "ground floor" of a building.—*Ground-plate* or *ground-sill*, the lowest horizontal timber into which the principal and other timbers of a wooden erection are inserted.

GROUNDS.—In joinery, pieces of wood attached to a wall for nailing the finishings to. They have their outer surface flush with the plastering.

GROUPED COLUMNS or PILASTERS. — A term used to denote three, four, or more columns or pilasters assembled on the same pedestal. When two only are placed together they are said to be *coupled*.

GUILLOCHE.—An interlaced ornament, formed by two or more intertwining bands, frequently used in classical architecture to enrich the torus and other mouldings.

Guttæ.

GUTTÆ.—Ornaments resembling drops, used in the Doric entablature, immediately under the triglyph and mutule.

GUTTER.—A channel of wood, iron, zinc, or lead, fixed to the eaves of a roof to receive the rain-water.— *Valley-gutter*, a sloping gutter (usually of boards covered with lead) formed in the valley of a roof.

H

HAFFIT, HAFT.—The fixed part of a lid or cover, to which the movable part is hinged; a handle.

HALF-LONG (Scotch, *Halflin*).—One of the bench-planes.

HALF-ROUND.—A moulding whose profile is a semicircle; a bead; a torus.

HALF-SPACE.—See FOOT-SPACE.

HALF-TIMBERED WORK. — A method of constructing walls with timber framework, the panels being filled with lath and plaster, bricks, &c.

HALVING. — A mode of joining two timbers by cutting a piece out of each to half the depth, and then overlapping the timbers.

HAMMER-BEAM.—A short beam attached to the foot of a principal rafter in a roof, in the place of the tie-beam, but not extending across the apartment. It is generally supported by a rib rising from a corbel below; and in its turn forms the support of

Hammer-beam Roof, Westminster Hall.

another rib, constituting with that springing from the opposite hammer-beam an arch.

HANDRAIL.—A rail to hold by. It is used in staircases to assist in ascending and descending. When it is next to the open newel it forms a coping to the stair balusters.

HANGING-BUTTRESS.—A buttress supported on a corbel.

HANGINGS.—Linings for rooms, consisting of tapestry, leather, paper, and the like.

HANGING STILE OF A DOOR OR GATE.—That to which the hinges are fixed.

HARDWOOD.—See SOFTWOOD.

HATCH.—A trap-door; a small opening in a wall or partition at some height above the floor, as a *serving-hatch*.

HATCHET.—A small axe with a short handle, used with one hand.

HAUNCH.—The wall or space above an arch between the vertex or crown and the springing; a small projection on the shoulder of a tenon.

HEADING-JOINT.—The joint of two or more boards at right angles to their fibres.

HEAD-PIECE.—The capping-piece of a partition, or of any series of upright timbers.

HEAD-POST.—The post in the stall-partition of a stable which is nearest to the manger. The post at the other end is the *heel-post*.

Hanging-Buttress.

HEART-WOOD.—The central part of the trunk of a tree; the *duramen*.

HELICAL LINE OF A HANDRAIL.—The spiral line twisting round the cylinder, representing the squared handrail before it is moulded.

HELIX.—A scroll or volute. The small volute or caulicule under the abacus of the Corinthian capital. In every perfect capital there are sixteen helices, two at each angle and two meeting under the middle of each face of the abacus.

HEMICYCLE.—A semicircle.

HEPTAGON.—A figure having seven sides and angles.

HERRING-BONE WORK.—Courses of stone or timber laid angularly, so that those in each course are placed obliquely to the right and left alternately. Wood-block and parquet floors are often laid in this way.

Helices.

HEXAGON.—A figure of six sides and six angles.

HINGES.—The joints on which doors, lids, gates, shutters, &c., are made to swing, fold, open, or shut up.

HINGING.—The operation of fixing hinges.

HIP.—The line of meeting of two inclined sides of a roof not in the same plane.—*Hip-knob*, a finial or other similar ornament placed on the top of the hip of a roof.—*Hip-moulding* or *hip-mould*, any moulding on the hip-rafter, or the *backing* of a hip-rafter.—*Hip-rafter*, the rafter which forms the hip of a roof; a *piend-rafter.*—*Hip-roof*, a roof the ends of which rise from the wall-plates, with the same inclination as the other two sides; called in Scotland a *piend-roof*.

HIT-AND-MISS VENTILATOR.—A kind of double grating with vertical bars of wood or metal, one grating being fixed and the other being free to slide so that its bars can be brought opposite either the bars or the interstices of the fixed grating.

HOARDING.—A temporary timber enclosure round a building; a timber structure for the display of advertisements.

HOGGING.—The drooping of the extremities and consequent convex appearance of any timber supported in the middle.

HOLDFAST.—A hook or long nail, with a flat short head, for securing objects to a wall; a bench-hook.

HOLING.—Piercing the holes for the rails of a stair, &c.

HOLLOW.—A concave moulding. Also called a *casement*.

HOLLOW NEWEL.—A well or opening in the middle of a staircase; the ends of the steps next the hollow are unsupported, the other ends only being supported by the surrounding wall of the staircase. The term is used in contradistinction to a solid newel, which has the end of the steps built into it.

HOOD-MOULDING, HOOD-MOULD.—The upper and projecting moulding over a Gothic door or window, &c.; called also a *label*, *drip*, or *weather-moulding*. See DRIPSTONE.

HOOK.—A "crook" or bent piece of iron secured to the joint of an opening, and forming a pivot on which the hinge of a door or gate turns.

HOUSE, *v.*—To excavate a space in one timber for the insertion of another.

HOUSING.—1. The space taken out of one solid to admit of the insertion of the extremity of another, for the purpose of connecting them.—2. A niche for a statue.

HYPERBOLA.—A section of a cone by a plane parallel to the axis.

HYPOTRACHELIUM.—The neck of the capital of a column; the part which forms the junction of the shaft with its capital.

I

ICOSAHEDRON.—A solid of twenty equal sides.

IMPOST.—The mouldings forming a cap or cornice to a pier, abutment, or pilaster, from which an arch springs.

INCISE.—To cut in; to carve.

INDENTED.—Cut into points like teeth, as an *indented* moulding.

INLAY.—In cabinet-making, to ornament a surface with thin slices of wood, ivory, pearl, &c., arranged in patterns. Inlaid work is also known as *marquetry*.

INSERTED COLUMN.—An *engaged column*.

INTAGLIO.—Literally, a cutting or engraving; hence, anything engraved, or a precious stone with arms or an inscription engraved on it.

INTERCOLUMNIATION.—The space between two columns.

b, Impost.

INTERLACING ARCHES.—Circular arches which intersect each other; frequent in arcades of the Norman style.

INTERTIE, INTERDUCE.—A piece of timber introduced horizontally between uprights, to bind them together or to stiffen them.

INTRADOS.—1. The interior or concave curve of an arch. See ARCH.—2. The concave surface of a vault.

IONIC ORDER.—The second of the three Grecian and third of the five Roman orders. The distinguishing characteristic of this order is the voluted capital of the column. The Ionic, as used by

Interlacing Arcade, Norwich Cathedral.

the Italian Renaissance architects, varied much in proportions; but the following are those given by Sir William Chambers:—The column is nine diameters high, of which the base occupies 30, and the capital 21 minutes. The architrave, divided into two fascias, is 40½ minutes, the frieze 40½, and the cornice 54.

ITALIAN ARCHITECTURE.—A term which is often used to designate the Renaissance architecture of Italy, of which there are three great schools—the Florentine, the Roman, and the Venetian.

J

JACK-PLANE.—One of the bench-planes, used in reducing inequalities in the timber preparatory to the use of the trying-plane.

JACK-TIMBERS.—Those timbers in a series which, being intercepted by some other piece, are shorter than the rest. Thus, in a hipped-roof, each rafter which is shorter than the side-rafters is a *jack-rafter*.

JACK-SCREW or LIFTING-JACK.—See SCREW-JACK.

JACOB'S LADDER.—A ladder consisting of a single plank pierced with holes for the feet, and fixed about 3 inches from the wall; chiefly used for access to lofts.

JAMB.—The vertical side of any aperture, such as a door, a window, or chimney.—*Jamb-linings*, the linings of the vertical sides of a doorway.—*Jamb-posts*, the upright timbers on each side of a doorway; called also *prick-posts*.

JERKIN-HEAD.—The end of a roof not hippel down to the level of the side walls, the gable being carried up higher than those walls. A truncated hipped roof, or half-hipped roof.

JERRY.—A contemptuous epithet denoting bad workmanship and material, as in *jerry-work*, *jerry-building*, &c.

JESTING-BEAM.—A beam introduced for the sake of appearance and not for use.

JIB-DOOR.—A door with its surface in the plane of the wall in which it is set, and intended to be concealed; it has no architraves, and the plinth and dado are carried across it.

JOGGLES, or JOGGLE-JOINTS.—In masonry, this term is applied to almost every sort of jointing in which one piece of stone is let or fitted into another, so as to prevent all sliding on the joints. In carpentry, the struts of a roof are said to be joggled into the truss-posts and into the rafters.

JOINT.—In carpentry and joinery, the place where one board or member is connected with another. Joints receive various names, according to their forms and uses.

JOINTER.—The largest plane used in straightening the edges of boards to be jointed together.

JOISTS.—The pieces of timber to which the boards of a floor or the laths of a ceiling are nailed.—*Trimming-joists*, two joists, into which each end of a small beam, called a *trimmer*, is framed. See TRIMMER.—*Binding-joists*, or *binders*, in a double floor, are those which form the principal support of the floor, and run from wall to wall.—*Bridging-joists*, those which are bridged on to the binding-joists, and carry the floor: they are laid across the binding-joists.—*Ceiling-joists*, joists to sustain a ceiling only.

JUMP.—An abrupt rise from a level course.

K

KERB or KIRB-PLATE.—See CURB-PLATE.

KERF.—The channel or way made through wood by a saw.

KEY.—A name given to all fixing wedges, and to the loose piece of metal by which a lock is operated. Also, a bar of wood grooved into a series of boards across the grain of the latter to prevent warping.

KEY-HOLE SAW.—A saw used for cutting out sharp curves, such as key-holes. It consists of a narrow blade, thickest on the cutting or serrated edge, its teeth having no twist or set, and a long handle perforated from end to end, into which the blade is thrust to a greater or less extent. Also called a *turning-saw*.

KEY-PILE.—The centre pile plank of one of the divisions of sheeting-piles contained between two gauge-piles of a cofferdam, or similar work. It is made of a wedge form, narrowest at the bottom, and, when driven, keys or wedges the whole together.

KEY-STONE.—The highest stone of an arch. See ARCH.

KILLESSE, CULLIS, COULISSE.—A gutter, groove, or channel. The term is corruptly applied in some districts to a hipped roof; as a *killessed* or *cullidged* roof. A dormer-window, too, is sometimes called a *killessed* or *cullidged* window.

KING-PIECE, KING-POST, KING-ROD.—The wood post or iron rod which, in a truss, extends between the apex of two inclined pieces and the tie-beam which unites their lower ends, as in a king-post roof.

KIOSK.—A Turkish word signifying an open pavilion or summer-house supported by pillars.

KNEE.—1. A piece of timber somewhat in the form of the human knee when bent.—2. A part of the back of a handrail which is of a convex form, the reverse of a *ramp*, which is concave.—*Knee-piece* or *knee-rafter*, an angular piece of timber used to strengthen the joining of two pieces of timber in a roof.

KNOCKER.—A kind of hammer fastened to a door, to be used in seeking for admittance.

KNOT or KNOB.—A bunch of leaves, flowers, or similar ornament, as the bosses at the ends of labels, the intersections of ribs, and the bunches of foliage in capitals.—*Knob*, a round, oval, or polygonal handle.—*Knot*, a hard part of wood at the junction of a branch with the stem.

Knocker, Village of Street, Somersetshire.

KNOTTING.—A process to prevent the knots of wood from appearing, by laying on a size composed of red-lead, white-lead, and oil, or a coat of gold-size, before painting.

KNUCKLE.—A joint of a cylindrical form, with a pin as axis, such as that by which the straps of a hinge are held together.

KYANIZE, *v.*—To steep in a solution of corrosive sublimate, as timber, to preserve it from dry-rot.

L

LABEL.—See DRIPSTONE.

LABYRINTH-FRET.—A fret with many involved turnings.

LACUNAR.—1. One of the coffers or sunk compartments in some varieties of ceilings.—2. A ceiling having sunk or hollowed compartments without spaces or bands between the panels; a lacquear having bands between the panels.

LADY-CHAPEL.—A chapel dedicated to the Virgin Mary, frequently attached to large churches.

LAGGINS, LAGGING.—The planking laid on the ribs of the centring of a tunnel or bridge, to carry the brick or stone work.

LAID-ON.—A term applied to mouldings that are got out in strips and nailed on to a surface.

LAMINATED ARCHES.—Arches composed of thin plates of wood fastened together.

LANCET ARCH.—One whose head is shaped like the point of

a lancet.—*Lancet Window*, a window with a lancet arch, characteristic of the Early English style of architecture.

LANDING.—The first part of a floor at the end of a flight of steps. Also, a resting-place between flights.

LANTERN.—1. An erection on the top of a dome or roof, to give light, and serve as a sort of crowning to the fabric (see LOUVRE). Also, the lower part of a tower placed at the junction of the cross in a cathedral or large church, having windows on all sides.—2. A top-light with glazed sides and often with glazed roof, raised above the flat or pitched roof of a room or corridor.

LAP.—The distance to which a board is laid over that below it; the distance to which a slate or tile is laid over the next but one below it.

Lancet Window, Comberton.

LAQUEAR.—See LACUNAR.

LARMIER.—The drip of a cornice; corruptly *lorimer*.

LATCH.—A fastening by which a door, &c., can be fastened when closed, without being locked or bolted.

LATH.—1. A thin narrow board or slip of wood nailed to the rafters of a building, to support the tiles or covering; a batten.—2. A thin narrow slip of wood nailed to the studs and ceiling-joists, to support the plastering.

LATHE.—A machine for turning wood or metal.

LATTICE.—Any work of wood or iron, made by crossing laths, rods, or bars, and forming open chequered or reticulated work; used in carpentry for the trusses of roofs and bridges.—*Lattice* or *lattice-window*, a window made of laths or strips of wood or iron which cross one another like net-work, so as to leave open interstices.

LAYER-BOARDS.—The boards for sustaining the lead of gutters.

LAY-PANELS.—Panels of greater width than height, in which the grain of the wood is horizontal.

LEAD NAILS.—Nails used to fasten lead, leather, canvas, &c., to wood. They are of the same form as clout-nails, but are covered with lead or solder.

LEAF.—One half of a double door; the flap of a table; the loose piece of "table-top" used in extending a dining-table.

LEAN-TO.—A building whose rafters pitch against or lean on to another building, or against a wall.

LEAR-BOARD.—The same as *layer board*.

LECTERN.—The reading-desk in the choir of churches.

LEDGE.—A surface projecting horizontally, or slightly inclined to the horizon; a string-course; also, the side of a rebate against which a door or shutter is stopped, or a projecting fillet serving the same purpose as a door-stop, or the fillet which confines a window-frame in its place.

LEDGED DOORS.—Doors formed of boards, with cross-pieces or ledges on the back to strengthen them.

LEDGERS.—The horizontal timbers in scaffolding, parallel to the wall, and serving to support the *putlogs*.

LICH-GATE, LYCH-GATE.—A shed over the gate of a churchyard, to rest the corpse under; called also a *corpse-gate*.

LINE.—A piece of cord or string; a sash-cord.

LINING.—The covering of the surface of any body with a thinner substance. The term is applied to coverings in the interior of a building, coverings on the exterior being properly termed *casings*. Linings of a door or window are the coverings of the jambs and soffit.—*Lining-out stuff*, drawing lines on a piece of board or plank so as to cut it into thinner pieces.

LINTEL.—A horizontal piece of timber, iron, or stone placed over an opening.

LIST, LISTEL.—A fillet moulding.

LOBBY.—A small hall or waiting-room; also, an enclosed space surrounding or communicating with one or more apartments, such as the boxes of a theatre.

LOCK.—1. An instrument composed of springs, wards, and bolts, used to fasten doors, drawers, chests, &c., and usually operated by a loose key. The principal varieties are: (1) the *plate-lock*, in a wood case; (2) the *rim-lock*, in a metal case—these two kinds being fixed on the backs of doors; (3) the *mortise-lock*, ordinary or upright, and concealed in a mortise cut in the edge of the door; (4) the *night-latch*, which locks automatically when the door is closed, but requires a key to open it from the outside; (5) the *padlock*, a removable lock with a hinged hook for attaching to a staple; and (6) the *cupboard-lock*, a small rim lock, often with a brass case. *Hook-bolt locks* are used for sliding-doors, &c.; *rebated mortise-locks* are used for doors with rebated edges, such as a pair of doors.—2. A basin or chamber in a canal, or at the entrance to a dock, with gates at each end.

LOCKER.—1. A small cupboard.—2. A small closet or recess,

frequently observed near an altar in Catholic churches, and intended as a depository for the sacred vessels, water, oil, &c.

LOCK-RAIL.—The middle rail of a door, to or opposite which the lock or fastening is fixed.—*Lock-stile*, the vertical piece of framing in a door or gate to which the lock is fixed.

LODGE.—1. A small house in a park, forest, or domain, subordinate to the mansion; a temporary habitation; a hut.—2. A small reservoir for storing water.

LOFT.—The place immediately under the roof of a building, when not used as an abode; as *hay-loft*. The gallery of a church is sometimes termed the *loft* in Scotland.

Louis-Quatorze Ornament.

Louis-Quinze Ornament.

LOG.—A large piece of round or unhewn timber; also, a large piece of squared timber.

LOGGIA.—An open gallery or arcade on an upper story of a building; also, a similar feature on a ground-floor.

LONG PLANE, or JOINTER. See JOINTER.

LOOP-HOLE.—A narrow opening in a wall.

LOUIS-QUATORZE, STYLE OF.—A style of ornament developed in France during the reign of Louis XIV. towards the end of the seventeenth century.

LOUIS-QUINZE, STYLE OF.—A variety of the Louis-Quatorze style of ornament which prevailed in France during the reign of Louis XV., in which the want of symmetry, which characterizes the Louis XIV. style, is carried to an extreme.

LOUVRE, LOOVER, LOVER, or LANTERN.—A dome or turret rising out of the roof of the hall in our ancient domestic edifices; formerly open at the sides, but now generally glazed. They were originally intended to allow the smoke to escape, when the fire was kindled on dogs in the middle of the room. The open windows in church-towers are called *louvre-windows*, and the boards or bars which are placed across them to exclude the rain are called *louvre-boards*, corruptly *luffer-boards*.

LUMBER.—In America, timber sawn or split for use.

Louvre, Abbot's Kitchen, Glastonbury.

LUNETTE.—An aperture in a concave ceiling for the admission of light.

LYING PANELS. See LAY-PANELS.

M

M-ROOF.—A kind of roof formed by the junction of two common roofs, with a valley between them.

MAIN COUPLE.—The trussed principal of a roof.

MALLEABLE IRON.—Cast iron annealed with charcoal, the operation giving to the metal some of the properties of wrought iron.

MANSARD ROOF.—A roof formed with an upper and lower set of rafters on each side, the latter being of steeper pitch than the former. It is called also a *curb-roof*.

MANTEL-PIECE.—The ornamental work around a fireplace.

MARGINS or MARGENTS.—The flat parts of stiles and rails in framed or panelled work. See DOUBLE-MARGINED DOORS.

MARQUETRY.—See INLAY.

MASK.—A piece of sculpture representing some grotesque form, in friezes, panels of doors, keys of arches, &c.

MASTIC.—A kind of cement made by mixing litharge with pulverized calcareous stones, sand, and linseed-oil. A bituminous composition in which wood-block flooring is laid.

MATCH-PLANES.—Planes in pairs, used in grooving and tonguing boards, one plane being used to form the groove, and the other to form the tongue. Such boards are said to be *matched*.

MAUSOLEUM.—A sepulchral chapel, or edifice erected for the reception of a monument, or to contain tombs.

MEANDER.—A complicated variety of the fret ornament.

MEDALLION. — A circular, oval, or sometimes square tablet, bearing on it objects represented in relief, or an inscription.

Meander.

MEDIÆVAL ARCHITECTURE.—A term properly applied to denote the architecture which prevailed throughout the middle ages, or from the fifth to the fifteenth century. In popular language, however, it is restricted to the Romanesque and Gothic styles, which prevailed from the eleventh to the fifteenth century.

MEMBER.—Any subordinate part of a building, order, or composition, as a frieze or cornice; and any subordinate part of these, as a corona, a cymatium, a fillet.

METOPE.—The space between the triglyphs of the Doric frieze.

METRE.—A French measure equal to 39·37 English inches.

MEZZANINE.—A story of small height introduced between two higher ones, or the upper of two stories in part of a building which together equal the height of the principal story in the other part.

MEZZO-RILIEVO.—Middle relief. See DEMI-RILIEVO.

MILLED LEAD.—Lead rolled out into sheets by machinery.

MINARET.—A slender, lofty turret rising by different stages or stories, surrounded by one or more projecting balconies, common in mosques in Mahometan countries.

MINSTER.—A monastery; an ecclesiastical convent or fraternity; but it is said originally to have been the church of a monastery; a cathedral church.

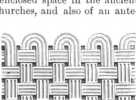

Minarets, Constantinople.

MITRE.—The line formed by the meeting of surfaces or solids at an angle. It is commonly applied, however, when the objects meet in a right angle, and the mitre-line bisecting this makes an angle of 45° with both.—Mitre-box, a box or trough with two parallel sides and a bottom, used for forming mitre-joints. It has cuts in its vertical sides, the plane passing through which crosses the box at an angle of 45°. The piece of wood to be mitred is laid in the box, and the saw being worked through the guide-cuts, forms the mitre-joint in the wood.—Mitre-square, a bevel with a fixed blade, for striking an angle of 45° on a piece of stuff.

MODILLION.—A block carved into the form of an enriched bracket, used under the corona of the Corinthian and Composite entablatures. Modillions less ornate are occasionally used in the Ionic entablature.

MONASTERY.—A house of religious retirement, whether an abbey, a priory, a nunnery, or a convent. The word is usually applied to the houses of monks.

MONKEY.—The ram or weight of a pile-driving engine.

Modillion.

MONTANTS, MOUNTINGS, MUNTINS.—The intermediate stiles in a piece of framing, which are tenoned into the rails.

MOORISH or MORESQUE ARCHITECTURE.—See SARACENIC.

MORTISE, MORTICE.—A cavity cut in a piece of wood or other material, to receive a corresponding projecting piece called a tenon, formed on another piece of wood, &c., in order to fix the two together at a right angle.—Mortise lock, a lock made to fit into a mortise cut in the stile of a door.

MOULDING.—A member, usually of curved section, formed on the edge or face of wood, stone, &c., for the purpose of ornament. In woodwork, mouldings are said to be stuck, when formed on the solid stuff, and planted or laid-in, when formed on separate pieces which are afterwards affixed to the framing.—Moulding-planes, joiners' planes used in forming the contours of mouldings.

MULLION, MUNNION, MONYCALE, MONIAL. — A vertical division between the lights of windows, screens, &c. Sometimes the stiles in wainscoting are called mullions.

MUTULE.—An ornament in the Doric cornice, answering to the modillion in the Corinthian, but differing from it in form, being a square block, from which guttæ depend.

N.

NAIL.—A small pointed piece of metal (usually with a head), to be driven into a board or other piece of timber, and serving to fasten it to other timber.

NAIL-HEAD MOULDING.—An ornament common in Norman architecture, consisting of a series of diamond-pointed knobs.

NAKED.—Any continuous surface, as opposed to the ornaments and projections which arise from it.—Naked flooring, the supporting timbers on which the floor-boards are laid.

NAOS.—The body of an ancient temple, sometimes, but erroneously, applied to the cella or interior. The space in front of the temple was called pronaos.

NARTHEX.—The name of an enclosed space in the ancient basilicas when used as Christian churches, and also of an ante-temple or vestibule without the church. Narthex is frequently used as synonymous with porch and portico.

NATTES.—A name given to an ornament used in the decoration of surfaces in the architecture of the twelfth century, from its resemblance to the interlaced withs of matting.

NAVE.—The central avenue or middle part of a church, extending from the western porch to the transept, or to the choir or chancel.

Nattes, Bayeux Cathedral.

NEBULE.—A moulding whose edge takes the form of an undulating line. It is used in corbel-tables and archivolts.

NECK, NECKING, or HYPOTRACHELIUM.—The part which connects a capital or head with its body or shaft; thus, the neck of a capital is the part between the lowest moulding of the capital and the highest moulding of the shaft. In the Grecian Doric it is the space between the annulets and channel, and in the Roman Doric the space between the annulets and astragal. The channels, astragals, or other members which terminate the shaft or body, are called the neck-mouldings. The neck of a finial is the part in which the finial joins the obelisk.

NEEDLE.—A beam of timber supported on upright posts, used to carry a wall temporarily during alterations or repairs.

NEEDLEWORK.—A term sometimes applied to the framework of timber, of which old houses are constructed.

NERVURES, NERVES, or BRANCHES.—The ribs which bound the sides of a groined compartment in a vaulted roof, as distinguished from the diagonal ribs.

NEUTRAL AXIS.—That plane in a beam in which theoretically the tensile and compressive forces terminate, and in which the stress is therefore nothing.

NEWEL.—The upright cylinder or pillar, round which, in a winding staircase, the steps turn; where there is no central pillar, the staircase is said to have an open newel. Also, the wood posts receiving the ends of handrails.

Niche, All-Souls' College, Oxford.

NICHE.—A recess in a wall for the reception of a statue, vase, or other ornament.

NOGGING.—Brickwork in panels between timber quarters.

NOGGING PIECES.—Horizontal pieces fitted in between and nailed to the quarters for strengthening the brick-nogging.

NOGS.—A north of England term for wood bricks.

NONAGON.—A figure having nine sides and nine angles.

NORMAN ARCHITECTURE.—A term commonly applied to the later Romanesque architecture of this country. It

Norman Door, Earl's-Barton, Northamptonshire.

was superseded, towards the end of the twelfth century, by the first of the Pointed or Gothic styles, the *Early English.* The Norman is readily distinguished by its general massive character, round-headed doors and windows, and peculiar enrichments.

NOSING.—The projecting edge of a moulding or board, such as a window-board, stair-tread, &c. It may be *square, rounded,* or *moulded.*

NOTCH, *v.*—To cut a hollow on the face of a piece of timber for the reception of another piece. *a* is the method of notching termed *halving; b* is a dovetailed notch; in *c* the notch is formed a little way from the end of each piece, so that the joint cannot be drawn asunder in either direction; in *d* the width of the notch is not so great as the width of the piece on which it is to be let down, which is also partially notched to receive it. This last is known as *caulking* or *cogging.*

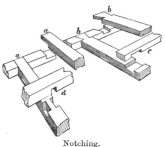

Notching.

NOTCH-BOARD.—A board notched or grooved to receive the ends of the boards which form the steps of a wooden stair.

O.

OBELISK.—A lofty, quadrangular, monolithic column of a pyramidal form; not, however, terminating in a point, nor truncated, but crowned by a flatter pyramid.

OBLIQUE ARCHES or OBLIQUE BRIDGES.—Those arches or bridges whose direction is not at right angles to their axes.

OCTAGON.—A figure of eight sides and eight angles.

OCTAHEDRON.—A regular solid contained by eight equal equilateral triangles.

OCULUS.—A round window; sometimes termed an O.

OFFSET or SET-OFF. — A horizontal break in a wall at a diminution in its thickness. In Scotland termed a *scarcement.* A curved or angular piece of rain-water pipe, &c., to fit a "break" in a wall.

OGEE.—See CYMA RECTA and CYMA REVERSA.

OGEE ARCH.—A pointed arch the sides of which are each formed with a double curve. It is used in the Decorated and Perpendicular styles, over doors, niches, tombs, and windows.

Ogee Arch.

OGIVE.—The French term for the ogee arch, but it is also applied to the diagonal ribs of a groined vault. The Pointed style of architecture is termed by the French *Le style Ogival.*

OILLETS, OYLETS.—In the walls of mediæval buildings, small openings or eyelet-holes through which missiles were discharged.

OPEN-NEWELLED STAIRS.—Stairs of two or more flights which have no solid pillar or newel in the centre.

OPEN-STRING.—A stair-string notched to receive the treads and risers.

ORATORY.—Originally, a small private chapel, or a closet near a bed-chamber, furnished with an altar, a crucifix, &c., and set apart for the purposes of private devotion.

ORB.—A plain circular boss. The mediæval name for the tracery of blank windows or stone panels.

ORDERS OF ARCHITECTURE.—The term applied to the varieties of Greek and Roman columnar architecture.

ORIEL WINDOW. — A large bay or recessed window. It usually projects from the outer face of the wall, either in a semi-octagonal or semi-square plan. Some writers restrict the term to projecting windows on an upper floor, supported on mouldings or corbels, and apply the term *bay-window* to such as rise from the ground.

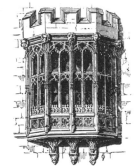

Oriel Window, Balliol College, Oxford.

ORIENTATION.—An eastern direction or aspect; the art of placing a church so as to have its chancel pointing to the east.

ORNAMENTS.—The smaller and detailed parts of the main work, not essential to it, but serving to adorn and enrich it.

OUTER STRING. See STRING-BOARD.

OVA.—Ornaments in the form of eggs, into which the ovolo moulding is often carved.

OVOLO.—A moulding, the vertical section of which is, in Roman architecture, a quarter of a circle; it is thence called the *quarter-round.* In Grecian architecture the section of the ovolo is elliptical, or rather egg-shaped.

P.

PACE.—A portion of a floor slightly raised above the general level; a dais.

PACKING.—A piece of wood inserted in framing, &c., to raise the whole or some portion to the required level.

PAD.—A handle.

PADDLE.—A small sluice.

PAGODA.—A native temple in the East Indies, China, &c.

PALE.—A sawn or cleft board fixed vertically and nailed to the rails of gates and fences.—*Paling,* a fence in which pales are used; if the pales touch each other or overlap, the fence is known as *close-pale* or *park-pale* fencing; if spaces intervene between the pales, it is an *open-pale* fence.

PAMPRES.—Ornaments consisting of vine leaves and grapes, with which the hollows of the circumvolutions of twisted columns are sometimes decorated.

PAN OF WOOD, or PAN OF FRAMING.—The panel formed by any three or more timbers, in half-timber work, &c.

PANEL.—In architecture, an area sunk from the general face of the surrounding work; also a compartment of a wainscot or ceiling, or of the surface of a wall, &c.; sometimes enclosing sculptured ornaments. In joinery, a thin piece of wood, framed in a groove by two upright stiles and two transverse rails.—*Panelling,* the operation of covering or ornamenting with panels; framed work containing panels.—*Panel-saw,* a saw used for cutting very thin wood; its blade is about 26 inches long, and it has about six teeth to the inch.

PANTILE or PENTILE.—A tile in the form of a parallelogram, straight in the direction of its length, but with a waved surface transversely.

Pantile.

PARABOLA.—A section of a cone by a plane parallel to one of the sides.

PARADISE. — In mediæval architecture, a small private apartment or study.

PARALLEL.—A line which throughout its length is equidistant from another line.

PARALLELOGRAM.—A quadrilateral figure of which the opposite sides are equal and parallel.

PARALLELOGRAM OF FORCES.—If two forces are represented in magnitude and direction by two lines meeting at a point, the diagonal of a parallelogram constructed on these lines will represent the resultant force.

PARAPET.—Literally, a wall or rampart to the breast, or breast-high, and serving as an ornament or for protection.

PARGET.—1. Gypsum; plaster-stone.—2. Plaster laid on roofs or walls.—3. A plaster formed of lime, hair, and cowdung, used for plastering flues.

PARING CHISEL.—A broad, flat chisel, used by joiners; it is worked by the impulsion of the hand alone, and not by the blows of a mallet, like the socket-chisel, firmer, &c.

PARQUETRY.—A species of flooring composed of small pieces of wood, from $\frac{1}{4}$ to 1 inch in thickness, arranged in patterns. Any inlaid woodwork of similar character.

PARTING BEAD. — The beaded slip inserted into the centre of the pulley-stile of a window, to keep apart the upper and lower sashes.

PARTITION.—A wall of stone, brick, or timber, which serves to divide one apartment from another in a building. Timber partitions are composed of sills, heads or capping pieces, studs or quarterings, posts, &c., and may be trussed.

Parquetry.

PARTY-WALLS.—A wall formed between houses to separate them from each other, and prevent the spreading of fire.

PATAND, PATIN.—A piece of timber laid on the ground to receive the ends of vertical pieces; a bottom plate; a sill.

PATERA.—1. An open vessel in the form of a cup, used by the Greeks and Romans in their sacrifices and libations.—2. The representation of a cup or round dish in flat relief, used as an ornament in friezes; but many flat ornaments are called pateras which have no resemblance to cups or dishes.

PATTEN.—The base of a column or pillar.

PAVILION.—1. A turret or small building, usually isolated, and having a tent-formed roof, whence its name.—2. A projecting part of a building, when it is carried higher than the general structure, and provided with a tent-formed roof.

PECKY.—A term in America applied to timber in which the first symptoms of decay appear.

PEDESTAL.—An insulated basement or support for a column, a statue, or a vase. It usually consists of a base, a die or dado, and a cornice, called also a *surbase* or *cap*. When a range of columns is supported on a continuous pedestal the latter is called a *podium* or *stylobate*.

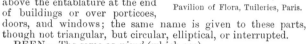
Pavilion of Flora, Tuileries, Paris.

PEDIMENT.—In classic architecture, the triangular finishing above the entablature at the end of buildings or over porticoes, doors, and windows; the same name is given to these parts, though not triangular, but circular, elliptical, or interrupted.

PEEN.—The same as *piend* (which see).

PENDANT.—A hanging ornament used in the vaults and timber roofs of Gothic architecture, and in plastered ceilings; a fitting for gas or electric light, suspended from a ceiling.

PENDANT POST.—1. In a mediæval roof-truss, a short post placed against the wall, its lower end supported on a corbel or capital, and its upper end carrying the tie-beam or hammer-beam.—2. The support of an arch across the angles of a square.

PENDENTIVE.—The portion of a dome-shaped vault which descends into a corner of an angular building, when a ceiling of this kind is placed over a straight-sided area. In Gothic architecture, the portion of a groined ceiling springing from one pillar or impost, and bounded by the apices of the longitudinal and transverse vaults.—*Pendentive bracketing* or *cradling*, the coved bracketing springing from the walls of a rectangular area in an upward direction, so as to form the horizontal plane into a circle or ellipse; the timber work for sustaining the lath and plaster in pendentives.

PENTAGON.—A figure with five equal sides and angles; if the sides are unequal, it is an irregular pentagon.

PENT-HOUSE.—A shed with a roof of a single slope.—*Pent-roof*, a roof of a single slope; a *shed-roof*.

PERCH.—An old name for a bracket or corbel.

PERCLOSE, PARCLOSE.—The raised carved timber back to a bench or seat; the parapet round a gallery; a screen or partition.

PERGOLA.—A light trellis in the form of an arch or small building, over which plants are trained.

PERIBOLUS.—A court surrounding a temple, and itself surrounded by a wall enclosing the whole of the sacred ground.

PERIDROMUS.—The space in a peripteral temple between the walls of the cella or body and the surrounding columns.

PERIMETER.—The circuit or boundary of any plane figure.

PERIPHERY.—The circumference of a circle or ellipse, or of any curvilinear figure.

PERIPTERAL.—A temple, the cella of which is surrounded with columns, those on the flanks being at a distance from the wall equal to their intercolumniation.

PERISTYLE, PERISTYLIUM.—A range of columns surrounding anything, as the cella of a temple, or any place, as a

Tower of Magdalen College, Oxford.

court or cloister. It is frequently but incorrectly limited in signification to a range of columns round the interior of a place.

PERPENDICULAR LINES.—Lines at right angles to other lines. Sometimes used as synonymous with *vertical lines*.

PERPENDICULAR STYLE.—The third and last of the Pointed or Gothic styles used in Britain. It was developed from the Decorated during the latter part of the fourteenth century, and continued in use till the middle of the sixteenth. The chief characteristics are the preponderance of perpendicular lines in the tracery of windows, the panelling of flat surfaces within and without, and the multiplicity of small shafts with which the piers, &c., are overlaid. Its magnificent church towers form a leading beauty of this style. In the tower of Magdalen College, Oxford, the decoration is reserved almost entirely for the uppermost story.

PERSIAN.—A figure in place of a column, used to support an entablature.

PIAZZA.—A square open space surrounded by buildings or colonnades. The term is frequently, but improperly, used to signify an arcaded or colonnaded walk.

PIECE-WORK.—Work done and paid for by the measure of quantity, in contradistinction to *day-work*, which is done and paid for by the measure of time.

PIEND.—An arris; a salient angle; a hip; the portion of a roof between or contiguous to hips, the hip-rafters being termed piend-rafters.

PIEND CHECK.—A term applied in Scotland to the rebate formed on the piend or angle at the bottom of the riser of the stone step of a stair, as at *a a a* in the figure.

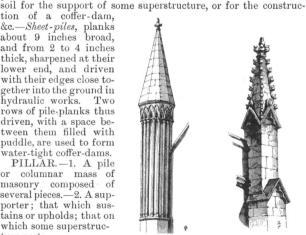
Piend Check.

PIER.—1. The support of the arches of a bridge; the solid parts between openings in a wall. (See ARCH.)—2. A mole or jetty carried out into the sea.—3. The pillars in Norman and Gothic architecture are generally, though not very correctly, termed *piers*.

PILASTER.—A debased pillar; a square pillar projecting from a pier or wall to the extent of from one-fourth to one-third of its breadth. Pilasters originated in the Grecian antæ.

PILE-DRIVER or PILE-ENGINE.—An engine for driving piles. In its simplest form it consists of a large ram or block of iron, termed the *monkey*, which slides between two guide-posts. Being drawn up to the top, and then let fall from a considerable height, it comes down on the head of the pile with a violent blow. It may be worked by men or horses, or a steam-engine.

Pile-driver.

PILES.—Beams of timber, pointed at the end, driven into the soil for the support of some superstructure, or for the construction of a coffer-dam, &c.—*Sheet-piles*, planks about 9 inches broad, and from 2 to 4 inches thick, sharpened at their lower end, and driven with their edges close together into the ground in hydraulic works. Two rows of pile-planks thus driven, with a space between them filled with puddle, are used to form water-tight coffer-dams.

PILLAR.—1. A pile or columnar mass of masonry composed of several pieces.—2. A supporter; that which sustains or upholds; that on which some superstructure rests.

PIN.—A piece of wood or metal, square or cylindrical in section, used to fasten timbers together; a

Early English Pinnacle, Beverley Minster. Perpendicular Pinnacle, Trinity Ch., Cambridge.

dowel or peg. Large metal pins are termed *bolts*, and the wooden pins used in shipbuilding *trenails*.

PINNACLE.—A turret, or part of a building elevated above the main building. In mediæval architecture, any smaller structure or ornament, consisting of a body or shaft terminated by a pyramid or spire, used either exteriorly or interiorly.

PINNING.—Fastening tiles or slates with pins; inserting small pieces of stone to fill vacuities.—*Pinning up*, driving in wedges in the process of under-pinning, so as to bring the upper work to bear fully on the work below.

PIPE-CASING.—A board screwed to grounds to afford access to pipes fixed on the wall behind.

PISCINA.—A niche on the south side of the altar in churches, containing a small basin and water-drain.

· PITCH OF A ROOF.—The inclination of the sloping sides of a roof to the horizon.

PITCHING-PIECE. See APRON-PIECE.

PLAFOND, PLATFOND.—The ceiling of a room, whether flat or arched. The under side of a cornice. Generally, any soffit.

PLAIN OR PLANE TILES.—Parallelograms of burnt clay used for covering roofs, &c., and generally about 10½ inches long, 6¼ inches wide, and ⅝ inch thick.

Piscina, Fifield, Essex

PLAN.—Anything drawn or represented on a plane, as a map or chart; but the word is usually applied to the horizontal geometrical section of anything, as a building, for example.

PLANCEER, PLANCHER.—A ceiling, or the soffit of a cornice.

PLANE.—A cutting instrument for wood, consisting of a chisel guided by the *stock* in which it is set.

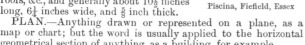

Jack-Plane.   Smoothing-Plane. Compass-Plane.

Rebate-Plane.   Filister, side and end.   Plough-Moulding.

PLANING MACHINE.—A power-machine for planing wood.

PLANK.—A piece of sawn timber, 11 or 12 inches wide, and from 1½ to 3 or 4 inches thick.

PLANTED.—In joinery, a small member wrought on a separate piece of stuff, and afterwards fixed in its place.

PLAT-BAND.—1. Any flat rectangular moulding, the projection of which is much less than its width; a fascia.—2. A lintel formed with voussoirs in the manner of an arch, but with the intrados horizontal.—3. The fillets between the flutes of the Ionic and Corinthian pillars.

PLATE.—A general name for all timber laid horizontally in a wall to receive the ends of other timbers, such as a *wall-plate*.

PLATE-RACK.—A series of narrow shelves to hold plates on edge.

PLATFORM.—A flat covering or roof of a building; a terrace or open walk on the top of a building; a raised part of a hall for the use of speakers, singers, and others.

PLETHORA.—A disease of trees.

PLINTH.—A square member serving as the base of a column, the base of a pedestal, or of a wall. See COLUMN.

PLOT, *v.*—To make a plan of anything from measurements.

PLOUGH.—A joiner's grooving-plane.

PLUG.—A small piece of wood driven into a hole in a wall, and serving as an attachment for a nail in fixing joinery, &c.

PLUG CENTRE-BIT.—A modified form of the ordinary centre-bit, in which the centre-point or pin is enlarged into a stout cylindrical plug, which may exactly fill a hole previously bored, and guide the tool in the process of cutting out a cylindrical countersink around the same.

PLUMB.—Vertical.—*Plumb-line*, *Plumb-rule*, a board with parallel edges, down the middle of which a line is drawn, and to the upper end of this line the end of a string is attached, carrying a piece of lead at its lower end. When the edge of the

board is applied to an object, the exact coincidence of the plumb-line with the line marked on the board indicates that the object is vertical. Sometimes another board is fixed across the lower end of the plumb-rule, having its lower edge at right angles to the line drawn on the other. In this case it becomes a level.

PLUMMET.—1. A long piece of lead attached to a line, used in sounding the depth of water.—2. A *plumb-line* or *plumb-rule*.

POCKET.—A hole in the pulley stile of a sash-window for inserting the balance-weights.

POD-AUGER.—A name given in some localities to an auger formed with a straight channel or groove. See AUGER.

PODIUM.—In architecture, a continuous pedestal; a stylobate; also, a projection which surrounded the arena of the ancient amphitheatre, where sat persons of distinction.

POINTED ARCHITECTURE.—See GOTHIC.

POLE-PLATE.—A sort of wall-plate laid on the ends of the tie-beams of a roof, to receive the rafters.

POLINGS.—Boards used to line the inside of a trench or tunnel during its construction, to prevent the falling of the earth or other loose material.

POLYCHROMY.—Coloured decoration.

POLYGON.—A plane figure with "many angles"; the term is usually restricted to figures with more than four sides or angles. A regular polygon is one in which all the sides and angles are equal.

POMEL, POMMEL.—A knob or ball used as a finial to the conical or dome-shaped roof of a turret, pavilion, &c.

POPPY HEAD.—An ornament carved on the raised ends of seats, benches, and pews in churches of the Perpendicular style.

PORCH.—An exterior appendage to a building, forming a covered approach or vestibule to a doorway. The porches in some of the older churches are of two stories, having an upper apartment, to which the name *parvise* is sometimes applied.

PORTAL.—1. The smaller of two gates at the entrance of a building.—2. A term formerly applied to a little square corner of a room, separated from the rest by a wainscot, and forming a short passage into a room.—3. A kind of arch over a door or gate, or the framework of the gate.

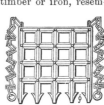

Poppy Head, Minster Church.

PORTCULLIS.—A strong grating of timber or iron, resembling a harrow, made to slide in vertical grooves in the jambs of the entrance gate of a fortified place. The portcullis formed an armorial bearing of the house of Lancaster, and is of frequent occurrence as a sculptured ornament on buildings erected by the monarchs of the Lancaster family, as on Henry VII's Chapel at Westminster Abbey.

PORTICO.—An open space before the entrance of a building, fronted with columns. Porticoes are distinguished as prostyle or in antis, as they project before or recede within the building. They are further distinguished by the number of their columns; as tetrastyle (with 4 columns), hexastyle (6), octastyle (8), decastyle (10).

Portcullis.

POST.—A piece of timber set upright, and intended to support something else; as the *posts* of a house, door, gate, fence. Any vertical piece of timber; as a *king-post*, *queen-post*, *truss-post*, *door-post*, &c.—*Post and railing*, a kind of open wooden fence, consisting of posts and rails, &c. See *Pale*, *Paling*.

POST AND PANE, POST AND PAN, POST AND PETRAIL.—Another name for walls composed of timber framing, with panels of brick, stone, or lath and plaster.

POSTERN.—Primarily, a back door or gate; a private entrance. Hence, any small door or gate.

PRICK-POST.—The same as *queen-post*.

PRINCESS-POSTS.—Subsidiary suspending posts in a truss, in which queen-posts are also used.

PRINCIPALS, or PRINCIPAL RAFTERS.—Those which are larger than the common rafters, and which are framed together with other timbers to form trusses. The struts, braces, &c., used in framing with the principal rafters are sometimes called *principal struts*, *principal braces*, &c.

PRISM.—A solid of which the ends are equal, similar, and parallel rectilineals, and the other sides are parallelograms.

PRISMOID.—A solid having for its two ends any dissimilar parallel plane figures of the same number of sides, and its upright sides trapezoids.

PROFILE.—The outline or contour of anything, such as a building, a figure, a moulding.

PROJECTION.—The jutting out of certain parts of a building beyond the naked wall, or of anything in advance of a normal line or surface. Also used to denote the plan (horizontal projection) and elevation (vertical projection) of any object.

PROSCENIUM.—That part of a theatre from the curtain or drop-scene to the orchestra. In the ancient theatre it comprised the whole of the stage.

PROTRACTOR.—An instrument for laying down and measuring angles on paper.

PUGGING.—Any composition, generally a coarse kind of mortar, laid in floors and partitions to prevent the transmission of sound. In Scotland it is termed *deafening*.

PUG-PILES.—Piles mortised into each other by a dove-tail joint. They are also called *dove-tailed piles.*

PULLEY.—A wheel over which a cord, chain, or belt is passed, to transmit motion, as in shafting, or to reduce friction, as in sash-pulleys, or to increase power, as in blocks and pulleys. Any wheel to reduce friction, as in the fittings of sliding doors.

PULLEY-MORTISE.—See Chase-Mortise.

PULLEY-STILE.—The stile of a window-case, to which the pulleys are fixed.

PULPIT.—An elevated place or enclosed stage in a church, in which the preacher stands. It is called also a *desk* or *rostrum*. Pulpits were also sometimes erected on the outside of churches.

PULVINATED.—A term used to express a swelling in any portion of an order, as in the modern Ionic frieze.

PUMP-BIT.—A species of large auger with removable shank, such as is commonly used for boring wooden pump-barrels.

PUNCHEON.—A short post; a small upright piece of timber in a partition, now called a *quarter*.

PURLIN.—A piece of timber laid horizontally, resting on the principals of a roof or on walls, to support the common rafters. Purlins are in some places called *ribs* or *wavers*.

PUTLOGS.—Short pieces of timber used in scaffolds to carry the floor. They are placed at right angles to the wall, one end resting on the ledgers of the scaffold and the other in holes left in the wall, called *putlog-holes*. See Ledger.

PUTTY.—A cement compounded of whiting and linseed-oil, and sometimes white-lead, used for stopping small cavities in wood-work and for fixing the glass in window frames.

PYRAMID.—A solid which has a rectilinear figure for its base, and for its sides triangles with a common vertex.

Q

QUADRA.—A square frame or border enclosing a bas-relief; any frame or border; the plinth of a podium.

QUADRANGLE.—A square surrounded with buildings.

QUADRILATERAL.—Four-sided.

QUARREL or Quarry.—A lozenge-shaped pane of glass; also, the opening in which the glass is set; a small square or lozenge-shaped paving tile or stone.

QUARTER-GRAIN.—When timber is split in the direction of its annular plates or rings. When it is split across these, towards the centre, it is called the *felt-grain*.

QUARTER-ROUND.—The ovolo moulding.

QUARTERS.—The vertical timbers of a partition to which the laths are nailed; called also *studs*, and in Scotland *standards*.

QUARTER-SPACE. See Foot-Space.

QUATREFOIL.—A piercing or panel formed by cusps or foliations into four leaves. An ornament similar to the four leaves of a cruciform flower, frequently used as a decoration in hollow mouldings in the Early English and Decorated styles.

Quatrefoils.

QUEEN-POST, Queen-Rod.—A suspending post or rod on either side of the centre of a truss.

QUINDECAGON. — A plane figure with fifteen sides.

QUIRK. — An indentation at the side of a convex moulding. A torus-moulding with a quirk on one side, as in skirting, is known as a single-quirk bead; with quirks on both sides, it is a double-quirk bead.

Quirked Ogee. Plain Ogee.

QUOIN.—The external angle of a building, and generally the stones of which that angle is formed.

R

RABBET. See Rebate.

RAD and Dab.—A substitute for brick-nogging in partitions, consisting of cob, or a mixture of clay and chopped straw, filled in between laths of split oak or hazel; called also *wattle and dab*.

RADIUS of a Circle.—Any straight line drawn from the centre to the circumference.

RAFTERS.—Pieces of timber which form the framework of the slopes of a roof. Common rafters are those to which the slate boarding or lathing is attached. See Principal Rafters.

RAGLET or Raglin (corruption of *regula*).—A groove cut in stone or brick work to receive lead flashing, &c.

RAGLINS.—A term used in the north of England for the slender ceiling joists of a building.

RAILS.—The horizontal timbers in any piece of framing.

RAISED PANELS.—Panels with the central portion thicker than the edges.

RAISING PLATE, Reson Plate.—The wall-plate, or more generally, any horizontal timber which, laid on walls or borne by posts or puncheons, sustains other timbers.

RAKE.—A slope or inclination.—*Raking mouldings*, those which are inclined from the horizontal line.

RAMP.—Literally, a spring or bound; any sudden rising interrupting the continuity of a line; a concave sweep connecting a higher and lower part of some work, as the coping of a wall, or the higher and lower parts of a stair-railing; a flight of steps, or the line tangential to the steps.—*Ramp and twist*, a line which rises and winds at the same time, as the handrail of winding stairs. —*Rampant arch*, one whose imposts are not of the same height.

RANCE.—A shore or prop. [Scotch.]

RAND.—A border or margin, or a fillet cut from a border or margin in the process of straightening it.

RASP.—A kind of coarse file used in finishing woodwork.

REBATE.—A rectangular longitudinal recess made in the edge of any substance. Thus the rectangular recess made in a door-frame, into which the door shuts, is a *rebate*. The rebated reveal of a door or window.—*Rebate-planes*, planes used in sinking rebates in joinery. Of these there are the moving fillister, used in sinking rebates on the edge of the board next to the workman, and the sash fillister in sinking the rebate on the edge farthest from him; and the guillaumes, skewed and square, the former for finishing the rebate across the direction of the fibre, and the latter for finishing it in the direction of the fibre.

RECESS.—A small cavity or niche in a vertical surface.

RECESSED ARCH.—One arch within another. Such arches are sometimes called double, triple, &c., or compound arches.

RECTANGLE.—A four-sided figure in which all the angles are right-angles.

REEDS.—A moulding consisting of beads side by side.

REGLET, Regula.—A small moulding, rectangular in its section; a fillet or listel. Also, a rectangular groove.

RELIEF, Rilievo.—The projection of a figure beyond the

High Relief. Low Relief.

ground or plane on which it is formed. Relief is of three kinds: high relief (*alto rilievo*), low relief (*basso rilievo*), and half relief (*mezzo rilievo*). The difference is in the degree of projection.

RENAISSANCE.—A term applied to the style of building and decoration which came into vogue in the early part of the sixteenth century, professedly a return to the classic architecture of Greece and Rome.

REREDOS.—The ornamental structure behind and above the altar in a church.

RESOLUTION and Composition of Forces.—The term *resolution of forces* signifies the dividing of any single force into two or more others, which, acting in different directions, shall produce the same effect as the given force. This is the reverse of *composition of forces*.

RESPOND.—A pilaster or half pillar responding to another similar, or to a whole pillar opposite to it.

RESTING POINTS.—In handrailing, the heights set up to obtain the section of a cylinder in forming the wreath.

RESULTANT.—In dynamics, the force which *results* from the *composition* of two or more forces acting upon a body.

RETICULATED MOULDING.—In architecture, a member composed of a fillet interlaced in various ways like net-work. It is seen chiefly in buildings in the Norman style.

RETURN.—A disease of trees. In building, a side or part that falls away from the front of any straight work.—*Return bead*, one which shows the same appearance on the face and edge of a piece of stuff, forming a double quirk.

REVEALS.—The sides of an opening for a door or window, between the framework and the face of the wall. In Scotland termed frequently *rybat-head* or *rebate-head*.

RHONE.—An eaves-trough or gutter. [Scottish.]

RIBBING.—An assemblage of ribs.

RIBS.—Purlins; also curved pieces of timber to which the laths are fastened, in forming domes, vaults, niches, &c. Projecting bands or mouldings used in vaulting, and in ornamented ceilings, both flat and curved.

RIDGE.—The highest part of the roof of a building, but more particularly the line of meeting of the upper ends of the rafters in a span-roof. Also the internal angle or nook at the crown of a vault.—*Ridge-piece*, *Ridge-plate*, a piece of timber at the ridge of a roof, against which the common rafters abut.—*Ridge-roll* or *Ridge-batten*, a rounded piece of timber, over which the lead is turned in the ridges and hips of a roof.—*Ridge-tile*, a convex tile made for covering the ridge of a roof.

RIPPING-SAW.—One used for cutting wood in the direction of the fibres.

RISER.—The vertical part of a step; also, the vertical framing under a seat, &c.

RISING HINGE.—One so constructed as to raise the door to which it is attached, as it opens.

ROCOCO.—A debased variety of the Louis-Quatorze style of ornament, proceeding from it through the degeneracy of the Louis-Quinze. The

Rococo Ornament.

term is sometimes applied in contempt to anything bad or tasteless in ornamental decoration.

ROD.—A measure of length equal to 16½ feet. A square rod is the usual measure of brickwork, and contains 272¼ square feet.

ROLL.—A piece of wood about 2 inches thick and rounded on the top, over which the lead or zinc covering of a flat roof is dressed.—*Roll-moulding*, a round moulding or large bead, some-

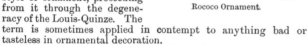

Roll-Moulding.   Roll-and-Fillet Moulding.

times divided longitudinally along the middle, the upper half of which projects over the lower. It occurs often in the later Early English and in the Decorated style.—*Roll-and-fillet moulding*, a round moulding with a square fillet on the face of it. It is most usual in the early Decorated style.

ROMAN ARCHITECTURE.—The style of architecture used by the Romans. It was founded on the Grecian architecture, but is further characterized by the use of the arch, vault, and dome.

ROMANESQUE.—A general term for all those round-arched styles of architecture which sprung from the Roman, and flourished in Western Europe till the introduction of Gothic architecture.

ROOD.—A cross, crucifix, or figure of Christ on the cross, placed in a church.—*Rood-loft*, the gallery in a church where the *rood* and its appendages were placed. This loft or gallery was commonly placed over the chancel-screen in parish churches, or between the nave and chancel; but in cathedral churches it was placed in other situations. The *rood tower* or *steeple* was that which stood over the intersection of the nave with the transepts.

ROOF.—The cover of a building, irrespective of the materials of which it is composed.—*Roof-timbers*, the timber framework of a roof, which may include common rafters, wall-plates, purlins, ridge-pieces, hip and valley rafters, and trusses.—*Roof-trusses*, combinations of timbers, including principal rafters, tie-beams, collar-beams, king-posts, queen-posts, struts, &c., framed together to form supports for the purlins, by which the common rafters are carried.

ROSE-WINDOW.—A circular window divided into compartments by mullions or tracery radiating or branching from a centre. It is called also *Catherine wheel* and *Marigold window*.

Rose-Window, St. David's.

ROSTRUM.—A scaffold or elevated place, where orations or pleadings are delivered; a pulpit.

ROTUNDA. — A round building; any building that is round both on the outside and inside.

ROUGH - BRACKETS. — Any unplaned pieces used as a groundwork for projecting features, such as plaster and wood cornices, coves, &c.

ROUGH-CAST.—An external wall-covering, composed of an almost fluid mixture of clean gravel and lime or cement, which is dashed on the wall previously prepared for its reception by a coating of cement-mortar, to which the rough-cast adheres.

ROUGH-HEW.—To hew coarsely without smoothing.

ROUGH - STRINGS.—Raking joists supporting a flight of stairs; known also as *carriages*.

ROUTTER-GAUGE. — A gauge used for cutting out the narrow channels intended to receive brass or coloured woods in inlaid work. It is formed like the common marking-gauge, but with a narrow chisel as a cutter, in place of the marking-point.

ROUTTER-PLANE.—A kind of plane used for working out the bottoms of rectangular cavities. The sole of the plane is broad, and carries a narrow cutter, which projects from it as far as the intended depth of the cavity.

RULE.—A straight-edge; a measuring instrument of boxwood, steel, &c., usually either 2 or 3 feet long.

S

SACRARIUM.—A sort of family chapel in the houses of the Romans, devoted to some particular divinity.

SACRISTY.—An apartment in a church where the sacred utensils are kept, and the vestments in which the clergyman officiates are deposited; now called the *vestry*.

SADDLE - BACKED COPING. — A coping thicker in the middle than at the edges; sloping both ways from the middle.

SAFE.—A tray, usually of wood covered with lead, fixed under baths, water-closets, cisterns, &c., and provided with a waste-pipe to carry off water leaking or overflowing from the fittings.

SAG.—To bend from a horizontal position.

SAIL-OVER, *v.*—To project.

SALIENT.—In architecture, a term used in respect of any projecting part or member.

SALLY.—A projection; also, the end of a piece of timber cut with an interior angle formed by two planes across the fibres, as the feet of common rafters; called in Scotland a *tace*.

SALOON.—A lofty spacious hall or apartment.

SAND-PAPER.—A stout paper covered on one side with powdered glass or quartz, and used for smoothing wood, &c. Also known as *glass-paper*.

SAP-WOOD. — The external part of the wood of exogens, through which the ascending fluids move most freely. For all building purposes the sap-wood is, or ought to be, removed from timber, as it soon decays.

SARACENIC ARCHITECTURE.—The architecture employed by the Saracens, who established their dominion over the greater part of the East in the seventh and eighth centuries. It may with equal propriety be styled Moslem or Mahometan architecture. One branch, the Moorish, is admirably exemplified in the architectural remains

Moorish Doorway, Cordova.

in Spain of the Moors or African Saracens, who subdued that country in the early part of the eighth century, and were only

finally driven from their last hold at Granada in 1492. The distinguishing characteristic is the profusion of colour and geometrical ornament with which the interior walls of buildings are overlaid.

SARCOPHAGUS.—A stone coffin or tomb.

SARKING.—The Scotch term for *slate boarding.* Such boarding is usually covered with sarking felt or waterproof paper.

SASH.—The framed part of a window in which the glass is fixed.—*Sash* (or *sashed*) *door,* a door with the upper part glazed like a window.—*Sash-fastener,* a latch or screw for fastening the sash of a window.—*Sash-frame,* the frame in which the sash is suspended, or to which it is hinged. When the sash is suspended, the frame is made hollow to contain the balancing weights, and is said to be *cased.*—*Sash-line,* the rope or chain by which a sash is suspended in its frame.—*Sash-saw,* a small saw used in cutting the tenons of sashes. Its plate is about 11 inches long, and has about 13 teeth to the inch. See DOUBLE-HUNG SASH.

SAW.—A cutting instrument consisting of a blade or thin plate of iron or steel, with one edge toothed. The *cross-cut saw* is for cutting logs transversely, and wrought by two persons, one at each end. The *pit-saw,* a long blade of steel with large teeth, and a transverse handle at each end, is used in saw-pits for sawing logs into planks or scantlings, and is wrought by two persons. The *frame-saw,* consisting of a blade from 5 to 7 feet long stretched tightly in a frame of wood, is used in a similar manner to the pit-saw. The *ripping-saw, half-ripper, hand-saw,* and *panel-saw* are saws for the use of one person, the blades tapering in width from the handle. *Tenon-saws, sash-saws, dovetail-saws,* &c., are saws made of very thin blades of steel, stiffened with stout pieces of brass, iron, or steel fixed on their back edges. They are used for forming the shoulders of tenons, dovetail-joints, &c., for which a clean cut is required. *Compass* and *key-hole saws* are long narrow saws, tapering from about 1 inch to ⅛ inch in width, and used for making curved cuts. Small *frame-saws* and *bow-saws,* in which very thin narrow blades are tightly stretched, are occasionally used for cutting both wood and metal. Nearly all heavy sawing is now done by machines, among which may be mentioned *log-frames, circular saws,* and *band-saws.*

SAXON ARCHITECTURE. — The Romanesque architecture which prevailed in England previous to the Norman Conquest.

SCAGLIOLA.—A plaster composition made to imitate marble.

SCALE.—A measuring rule used in preparing geometrical drawings of a different size from the object delineated.

a, Scamillus.

SCAMILLUS, SCAMILLI.—In ancient architecture, a sort of second plinth or block under statues, columns, &c., to raise them, but not, like pedestals, ornamented with any kind of moulding.

SCANTLING.—In carpentry, the dimensions of a piece of timber in breadth and thickness; also, a general name for small timbers, such as the quartering for a partition, rafters, purlins, or pole-plates in a roof, &c.

SCAPE, SCAPEMENT. See APOPHYGEE.

SCARFING. — A mode of lengthening beams, employed when it is necessary to maintain the same depth and width of the beam throughout. In doing this a part of the thickness of the timber of the length of the joint is cut from each beam, but on the opposite sides, so that they may lap on each other, and the parts, when united, are bolted or hooped together.

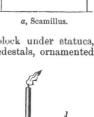

Sconce.

SCHEME ARCH, or SKENE ARCH.—A segmental arch.

SCONCE.—A branch to set a light upon, or to support a candlestick; a screen or partition to cover or protect anything; the head or top of anything. The term is sometimes used as synonymous with *squinch.*

Scotia or Trochilus Moulding.

SCONCHEON (from the French *écoinçon*).—A term applied to the portion of the side of the aperture of a door or window, from the back of the jamb or reveal to the interior of the wall.

SCOTIA.—The hollow moulding in the attic base between the fillets of the tori, or any similar moulding. It is sometimes called a *casement,* and often, from its resemblance to a pulley, *trochilus.*

SCREEN.—A kind of wire sieve for sifting sand, lime, gravel, &c. A partition of wood, wood and glass, &c., properly of less height than the room in which it is placed.

SCREW-JACK.—A portable machine for raising great weights by the agency of a screw.

SCRIBE.—A spike or large nail ground to a sharp point, to mark bricks on the face and back by the tapering edges of a mould, for the purpose of cutting them and reducing them to the proper taper for gauged arches.—*v.* To mark by a rule or compasses; to fit one piece to another, as one moulding against the face of another.

SCROLL.—An ornament resembling a narrow band arranged in convolutions or undulations.

SCROLL-SAW.—A relatively thin and narrow-bladed reciprocating saw, which passes through a hole in the work-table and saws a kerf in the work, which is moved about in any required direction on the table. The saw follows a scroll or other ornament, according to a pattern traced upon the work.

SEALING.—The operation of fixing a piece of wood or iron on a wall with plaster, mortar, cement, lead, or other binding.

SEASONING OF TIMBER.—The operation of drying timber, either naturally or artificially.

SEAT.—Something intended for sitting on; a bench; that part of a structure which is covered by another member resting upon it; the horizontal projection of an object.

SECTION.—In architecture, the projection or geometrical representation of a building, or part of a building, supposed to be cut by a vertical plane.

SECTROID. — The curved surface between two adjacent groins.

SEDILIA.—The Latin name for a seat, now generally applied to the seats for the priests in the south wall of the choir or chancel of churches and cathedrals.

SEGMENT.—A part cut off from anything, more especially a part cut off a circle by a straight line; the area contained by the arc of a circle and its chord, the chord being called the base of the segment, and the height of the arc the height of the segment.

Sedilia, Bolton Percy, Yorkshire.

SEGMENTAL ARCH.—An arch forming the segment of a circle, and less than a semicircle.

SET-OFF.—See OFFSET.

SET SQUARES.—Small triangles of celluloid, vulcanite, wood, &c., used by the draughtsman in drawing lines to form certain angles with the T-square, &c.

SETTING-OUT ROD.—A rod used by joiners for setting-out work, as windows, doors, stairs, &c.

SETTLEMENT.—Failure in a building occasioned by sinking.

SEVEREY, SEVERY, SEBEREE, SIBARY.—A compartment in a vaulted roof; also, a compartment or division of scaffolding.

SHAFT.—The shaft of a column is the body of it, between the base and the capital.—*Vaulting-shafts,* those which support ribs, or other parts of a vault.—*Shaft of a king-post,* the part between the joggles.—*Shaft of a chimney,* the part which rises above the roof for discharging the smoke into the air.

SHAKE.—A fissure or rent in timber, occasioned by its being dried too suddenly, or exposed to too great heat. Shakes frequently occur in growing timber from various causes.

SHANTY.—A hut or mean dwelling.

SHED ROOF.—The simplest kind of roof, formed by rafters sloping between a high and a low wall. Also, a roof of two unequal slopes meeting at a ridge, the steeper slope being glazed, —commonly used for weaving-sheds.

SHEERS.—Two masts or spars lashed or bolted together at the head, provided with a pulley, and raised to nearly a vertical position, used in lifting stones and other building materials.

SHEET-PILES.—See PILES.

SHELF.—A horizontal piece of boarding, slate, &c., for supporting articles of various kinds.

SHELL-BIT.—A boring tool used with the brace in boring wood; it is shaped like a gouge, that is, its section is the segment of a circle, and when used it shears the fibres round the margin of the hole, and removes the wood almost as a solid core.

SHINGLE.—A small piece of thin wood, used for covering a roof or building. Shingles are from 8 to 12 inches long and about 4 inches broad, and are laid to overlap like slates.

SHOE.—1. The inclined piece at the bottom of a water-trunk or lead pipe, for turning the course of the water and discharging it from the wall of a building. —2. An iron socket used in timber framing to receive the foot of a rafter or the end of a strut, &c.— 3. A pointed iron socket fixed on the driving end of a pile.

SHOOT.—To plane straight, or fit by planing.

SHOOTING-BOARD.—An external *fence* or guide used in shooting or planing the edges of boards, in which the piece to be planed is narrower than the face of the plane. The annexed figures are sections of shooting-boards, fig. 1 being used for a rectangular joint, and fig. 2 for a mitre joint. In both figures, *a* is a piece of board on which the

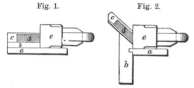

Fig. 1.   Fig. 2.

plane *e* lies on its side, and *b*, another piece on which the board to be planed, *d*, is laid; *c* is a stop against which the edge of the wood is pressed.

SHORE.—A piece of timber or other material placed in such a manner as to prop up a wall or other heavy body.—*Dead-shore*, an upright piece fixed in a wall that has been cut or broken through for the purpose of making some alterations in the building.—*Raking-shore*, an inclined shore.

SHOULDER.—Among artificers, a horizontal or rectangular projection from the body of a thing.—*Shoulder of a tenon*, the plane (generally transverse to the length of the piece of timber) from which the tenon projects.

SHREDDINGS.—In old buildings, short, light pieces of timber, fixed as bearers below the roof, forming a straight line with the upper side of the rafters.

SHRINE.—1. A reliquary, or box for holding the bones or other remains of departed saints, and usually like a small church with a high-ridged roof.—2. A tomb, of shrine-like configuration; and—3. A mausoleum of a saint, of any form.

SHUTTERS.—The boards or doors which cover a window. They are usually in two divisions, each subdivided into others, so that they may be received within the boxings into which they are folded back. The front shutter is of the exact breadth of the boxing, and also flush with it; the others are somewhat less in breadth, and are termed *back-folds* or *back-flaps*. Sometimes, in place of being hinged to fold back, they are suspended and counterbalanced like window-sashes, so as to slide.—*Revolving shutters* are made of laths jointed together and wound round a roller, which is usually placed horizontally above the soffit of the opening.

SIDE-HOOK.—A rectangular piece of wood, with a projecting knob at the ends of its opposite sides, used to hold a board fast, with its fibres in the direction of the length of the bench, while the workman is cutting across the fibres with a saw or grooving-plane, or is *traversing* the wood, which is planing it in a direction perpendicular to the fibres.

SIDE-TIMBERS, SIDE-WAVERS.—Local names for *purlins*.

SILICATE COTTON.—A fire-proof mineral fibre resembling wool, and used in floors, partitions, roofs, &c., to render them less inflammable, and to prevent extremes of temperature and the passage of sound.

SILL.—The horizontal piece of timber or stone at the bottom of a framed case, such as that of a door or window.—*Ground-sills* are the timbers on the ground which support the posts and superstructure of a timber building.—The word *sill* is also used to denote the bottom pieces which support quarter and truss partitions, &c.

SINGLE FLOOR, SINGLE FLOORING, SINGLE JOISTS, SINGLE-JOIST FLOOR.—Applied to naked flooring, consisting of bridging-joists only.

SINGLE-HUNG.—Applied to a window with two sashes, when only one is movable. When both sashes are movable, the window is said to be *double-hung*.

SINK.—A fixed trough-shaped vessel, in which kitchen utensils, &c., are washed; it may be of wood, or of wood lined with lead or copper, or of other materials.

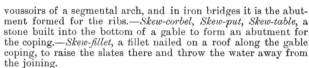

Sinks.

SITE.—The place whereon a building stands.

SKELETON.—Rough framing; a *skeleton-arch* is a rough arched framework of wood intended to be concealed by lath and plaster.

SKETCH.—An outline or general delineation of anything; a first rough or incomplete draught of a plan or design.

SKEW.—A term used in Scotland for a gable-coping or factable.—*Skew* or *Askew*, oblique; as a *skew*-bridge, *skew*-arch, &c.—*Skew-back*, the sloping abutment in brickwork or masonry for the ends of the arched head of an aperture. In bridges, it is the course of masonry forming the abutment for the

voussoirs of a segmental arch, and in iron bridges it is the abutment formed for the ribs.—*Skew-corbel, Skew-put, Skew-table*, a stone built into the bottom of a gable to form an abutment for the coping.—*Skew-fillet*, a fillet nailed on a roof along the gable coping, to raise the slates there and throw the water away from the joining.

SKIRTING, SKIRTING-BOARD.—The vertical board fixed to the walls round the margin of a floor, &c.

SKY-LIGHT.—A window placed in the top of a house, or a frame consisting of one or more inclined panes of glass placed in a roof to light passages or rooms below.

SLABS.—The outside planks or boards, mainly of sap-wood, sawn from the sides of round timber.

SLACK-BLOCKS.—The wedges on which the centres used in the construction of bridges are supported.

SLAG-WOOL.—See SILICATE COTTON.

SLAP-BOARDING, CLAP-BOARDING.—Feather-edged boarding; also boarding of irregular widths fixed to the sides of timber buildings.

SLAP-DASH.—A provincial term for *rough-cast*.

SLATE-BOARDING.—Close boarding covering the rafters of a roof, on which the slates are laid. In Scotland, called *sarking*.

SLEEPERS.—Pieces of timber on which are laid the ground joists of a floor; also the ground joists themselves. Formerly the term was used to denote the valley-rafters of a roof.—In railways sleepers are beams of wood or blocks of stone firmly imbedded in the ground to sustain the rails.

SLIDING-RULE.—A mathematical instrument or scale, consisting of two parts, one of which slides along the other, and each having certain sets of numbers engraved on it, so arranged that when a given number on the one scale is brought to coincide with a given number on the other, the product or some other function of the two numbers is obtained by inspection. Special slide-rules are made to suit the carpenter, engineer, gauger, &c.

SLIP-FEATHER.—A loose tongue of wood or iron for uniting the grooved edges of two contiguous boards, &c.

SLIT-DEAL.—Fir boards a full half-inch thick.

SMOOTHING-PLANE.—A small plane used in finishing joinery.

SNIPE'S-BILL PLANE.—A plane with a sharp arris for forming the quirks of mouldings.

SNOW-BOARDS.—Narrow boards with open joints nailed to battens, and laid over lead gutters, flats, &c., to prevent the blocking of the gutters and outlets.

SOCKET-CHISEL.—A chisel with a socket; a strong chisel used by carpenters for mortising, and worked with a mallet.

SOCLE.—A square member of less height than its horizontal dimension, serving to raise pedestals, or to support vases or other ornaments. It differs from a pedestal in being without base or capital. A *continued socle* is one continued round a building.

SOFFIT.—The under side of the lintel or ceiling of an opening; the lower surface of a vault or arch, or of the architrave or corona of an entablature.

SOFFIT-LINING.—The wood lining forming the soffit of a window, door, &c.

SOFTWOOD.—The timber of narrow-leaved coniferous trees, whether deciduous or not, such as pines, spruce, larch. The timber of broad-leaved trees is technically *hardwood*.

SOLE.—A sill; the part of a beam which rests on the support.—*Sole-plate*, a wall-plate.

SOUND-BOARDING.—Short boards, disposed transversely between floor-joists, and supported by fillets fixed to the sides of the joists, for holding the substance called pugging, intended to prevent sound from being transmitted from one story to another.

SOUNDING-BOARD or SOUND-BOARD.—A board or structure placed over a pulpit or rostrum, to reflect the sound of the speaker's voice, and thereby render it more audible.

SPAN.—In architecture, an imaginary line across the opening of an arch or roof, by which its extent is estimated.

SPANDREL.—The irregular triangular space comprehended between the outer curve or extrados of an arch, a horizontal line drawn from its apex, and a perpendicular line from its springing. In Gothic architecture, spandrels are often ornamented with tracery, foliage, &c. (See illustration of *Dripstone*.) Also, the space below the outer string of a flight of stairs.

SPANDREL-BRACKETING.—A cradling of brackets which is placed between curves, each of which is in a vertical plane, and in the circumference of a circle whose plane is horizontal.

SPAN-PIECE.—The collar-beam of a roof.

SPAN-ROOF.—Roofing formed by two inclined planes or sides, in contradistinction to a *shed* or lean-to.

SPAR.—A small beam or rafter. A common rafter of a roof, as distinguished from the principal rafters,

SPECIFICATION.—A statement of particulars, describing the manner of executing any work, and the quality, dimensions, and peculiarities of the materials to be used.

SPHERE, or GLOBE.—A solid bounded by a curved surface, every point of which is equidistant from a point called the centre.

SPHERICAL LATHE.—A lathe for turning spheres.

SPHEROID.—A body or figure approaching to a sphere, but not perfectly spherical. In geometry, a spheroid is a solid, generated by the revolution of an ellipse about one of its axes.

SPIKE.—A large nail; a drift-bolt.

SPIRE.—The pyramidal or conical termination of a tower or turret. The term is sometimes restricted to such tapering structures, crowning towers or turrets, as have parapets at their base. When the spire rises from the exterior of the wall of the tower without the intervention of a parapet, it is called a *broach-spire*.

SPIRE-LIGHTS.—The windows of a spire.

SPIRIT-LEVEL.—An instrument employed for determining a line or plane parallel to the horizon. It consists of a tube of glass nearly filled with spirit of wine or distilled water, hermetically sealed at both ends, and placed within a brass or wooden case, having a long opening on the side which is to be uppermost. When the instrument is laid on a horizontal surface, the air-bubble stands in the very middle of the tube; when the surface slopes, the bubble rises to the higher end.

SPLAY.—A sloped surface, or a surface which makes an oblique angle with another; as when the opening through a wall for a door, window, &c., widens inwards. A large chamfer.

SPOKE-SHAVE.—A sort of small plane used for dressing the spokes of wheels and other curved work, where the common plane cannot be applied.

SPOON-BIT.—A hollow bit with taper point for boring wood.

SPOUT.—An eaves-trough or gutter; a *down-spout* is a rain-water pipe.

SPRING BEVEL OF A RAIL.—The angle which the top of the plank makes with a vertical plane which has its termination in the concave side, and touches the ends of the rail-piece.

SPRINGER.—The point where the vertical support of an arch terminates and the curve begins. The lowest voussoir of an arch. The bottom stone of the coping of a gable.

SPRINGING.—The point from which an arch springs or rises. In carpentry, in boarding a roof, the setting the boards together with bevel joints, for the purpose of keeping out the rain.

SPROCKET.—See CHANTLATE.

SPUR.—Often used as a synonym for *strut*.

SQUARE.—1. A rectangle with four equal sides.—2. A T-shaped instrument used by the draughtsman, and known as a *T-square*.—3. An L-shaped instrument, usually of steel, divided into inches, &c., and known as the *carpenter's square*; it is used for setting out right angles and in solving various problems.—4. A measure of area, containing 100 square feet.

SQUARE FRAMED.—Panelled work in which all the angles of the stiles, rails, and muntins are square, *i.e.* not moulded.

SQUARE SHOOT.—A wooden trough for discharging water from a building.

SQUARE STAFF.—See ANGLE-BEAD, ANGLE-STAFF.

SQUARING A HANDRAIL.—The method of cutting a plank for a handrail, so that all the vertical sections may be rectangular.—*Squaring timber*, the operation of reducing a round log to a nearly square section, either by hewing with an adze, or by sawing off segments known as *slabs*.

SQUINCH.—The small pendentive arch formed across the angle of a square tower, to support the side of a superimposed octagon. A corner cupboard is also called a *squinch* or *sconce*.

SQUINT.—In mediæval architecture, a name given to an oblique opening in the wall of a church, generally so placed as to afford a view of the high altar from the transept or aisles.

STACK.—A pile of timber containing 108 cubic feet.—A group of smoke flues.—A series of rain-water pipes, &c.

STAFF-ANGLE, STAFF-BEAD.—See ANGLE-BEAD.

STAGE.—The part between one sloping projection and another in a Gothic buttress. Also, the horizontal division of a window separated by transoms. Sometimes the term is used to signify a floor, a story.—The part of a theatre, &c., in which the performance takes place.—A temporary floor in scaffolding, &c.

STAGGERED.—A term applied to a series of bolts arranged in zigzag form.

Squinch, Maxstoke Priory, Warwickshire.

STAIR.—A step, but generally used in the plural to signify a succession of steps, arranged as a way between two points at different heights in a building, &c. A succession of steps in a continuous line is called a *flight of stairs*; the termination of the flight is called a *landing*. Stairs are further distinguished by the various epithets, dog-legged, newelled, open newelled, &c.

STAIRCASE.—The building or apartment which contains the stairs; also the stairs themselves.

STAKE-FALD HOLES.—A local name for putlog holes.

STALLS.—Fixed seats, enclosed either wholly or partially at the back and sides, particularly those in the choir or chancel of a cathedral or church. In stables, partly-enclosed spaces for horses; larger spaces enclosed on all sides are called *loose boxes*.

STANCHEON.—A prop or piece of timber giving support to one of the main parts of a roof; any support composed of rolled iron or steel; also, one of the upright bars of a window, screen, railing, &c.

STANDARD. In joinery, any upright in a framing, as the quarters of partitions, and the frame of a door. A measure of sawn timber, that most commonly used being the Petersburg standard of 165 cubic feet.

Stalls, Higham Ferrers Church, Northamptonshire.

STAPLE.—A bent piece of metal, usually U-shaped, the two ends being pointed and driven into a wall or into wood to serve as a fastening for a padlock, bolt, &c.

STARLINGS or STERLINGS.—An assemblage of piles driven round the piers of a bridge to give them support.

STAY.—A brace, strut, or other support; a metal bar for holding a casement window or sky-light when open.

STEP.—One of the gradients in a stair; it is composed of two fronts, one horizontal, called the *tread*, and one vertical, called the *riser*. Also applied to other features of similar form.

STEREOGRAPHY.—The act or art of delineating the forms of solid bodies on a plane; a branch of solid geometry.

STEREOTOMY.—A branch of stereography, which teaches the manner of making sections of solids under certain specified conditions.

STICKING.—The operation of forming mouldings by means of a plane, in distinction from the operation of forming them by the hand. Also, the operation of forming mouldings on the solid framing, in distinction from mouldings formed on separate pieces and afterwards laid or planted in.

STILE.—The outermost vertical pieces forming the framing of a door, window, &c.; also, any vertical piece in framed work; horizontal pieces are known as *rails*.

STILTED ARCH.—An arch which does not spring immediately from the imposts, but is raised as it were upon stilts for some distance above them.

Stilted Arch.

STIRRUP PIECE.—A name given to a piece of wood or iron in framing, by which any part is suspended; a vertical or inclined tie.

STOCK AND BITS.—See BRACE AND BITS.

STOOTHING, STUDDING.—A provincial term for *battening*, and for *quartered partitions*.

STOPS.—Pieces of wood nailed on the frame of a door to form the recess or rebate into which the door shuts. Also, the slopes or ornaments at the end of a mould or chamfer. Also, an iron or other fitting, &c., against which a gate is stopped in closing.

STORY.—A stage or floor of a building, called in Scotland a *flat*; or a set of rooms on the same floor or level. In the United States, the floor next the ground is the first *story*; in France and England this is called the ground-floor, and the story above this is called the first floor or *story*.

STORY-POSTS.—Upright posts to support a floor or wall, through the medium of a beam placed over them.

STORY-ROD.—A rod used in setting up a staircase, equal in length to the height of a story of the building, and divided into as many parts as there are steps in the stair.

STOUP.—A basin for holy water, usually placed in a niche at the entrance of Roman Catholic churches. See illustration on p. 434.

STRAIGHT ARCH.—The term is usually and properly

applied to an arch over an aperture in which the intrados is straight. An arch consisting of two straight lines, meeting at the top, is also called a *straight arch.*

STRAIGHT-EDGE. — A slip of wood or iron made perfectly straight on the edge, and used to ascertain whether other edges or surfaces are straight.

Stoup. Maidstone Church, Kent.

STRAIGHT JOINT.—1. A joint which does not curve or depart from a straight line.—2. A term applied to the junction line of flooring boards when the joints at the butting ends of the boards form a continuous line.

STRAINING PIECE.—A beam placed between two opposite beams to prevent their nearer approach, as rafters, braces, struts, &c. If such a piece performs also the office of a sill, it is called a *straining sill.*

STRAP.—1. An iron plate placed across the junction of two timbers for the purpose of securing them together.—2. A batten on which the laths of interior walls are nailed.

STRETCHED OUT.—In architecture, a term applied to a surface that will just cover a body so extended that all its parts are in a plane, or may be made to coincide with a plane.

STRIKE-BLOCK.—A plane shorter than a jointer, used for shooting a short joint.

STRIKING.—In architecture, the drawing of lines on the surface of a body; the drawing of lines on the face of a piece of stuff for mortises, and cutting the shoulders of tenons. In joinery, the act of running a moulding with a plane.— *The striking of a centre* is the removal of the timber framing upon which an arch is built, after its completion.

STRING-BOARD, String-Piece, or Stringer.—A board supporting the ends of the treads and risers in wooden stairs, and known as *wall-string* or *outer string* according to its position, and as a *close, open* or *cut,* or *cut and mitred* string according to its form. A *bracketed* string is a cut string with ornamental brackets planted on under the ends of the treads.

STRING-COURSE.—A narrow moulding or projecting course continued horizontally along the face of a building, frequently under windows. It is sometimes merely a flat band.

STRINGING.—A narrow band in an inlaid or veneered panel.

STRUT.—Any piece of timber in a system of framing which is pressed in the direction of its length.—In flooring, short pieces of timber about 1½ inch thick and 3 to 4 inches wide, inserted between flooring joists (sometimes diagonally) to stiffen them.

STRUTTING-BEAM, Strut-Beam. — An old name for a *collar-beam.*

STRUTTING-PIECE.—The same as *straining-piece.*

STUB-MORTISE.—A mortise which does not pass through the whole thickness of the timber.—*Stub-tenon, Stump-tenon,* a tenon which does not pass through the whole thickness of the stuff in which it is mortised.

STUCCO.—A word applied as a general term to plaster of any kind, used as a coating for walls, and to give them a finished surface, but more particularly to external cement-work.

STUCK MOULDINGS.—See Sticking.

STUDS.—In carpentry, posts or quarters placed in partitions, usually about a foot distant from each other.

STYLOBATE.—Any sort of basement upon which columns are placed to raise them above the level of the ground or floor; more properly, a continuous unbroken pedestal, upon which an entire range of columns stands.

SUB-PLINTH. — A second and lower plinth placed under the principal one in columns and pedestals.

SUBSELLIA. — The small shelving seats in the stalls of churches or cathedrals, made to turn up upon hinges. They are also called *misereres.*

Subsellia, All-Souls, Oxford, the seat turned up.

SUMMER.—1. A large stone, the first that is laid over columns and pilasters. The first voussoir of an arch above the impost.— 2. A large timber serving as a lintel to an opening; a girder.

SUMMER-HOUSE.—A small building, often of rustic character, erected in gardens.

SUNK.—A term applied to work recessed below the general surface of the timber, &c.

SURBASE.—The crowning moulding or cornice of a pedestal; a border or moulding above the base, as the mouldings immediately above the base or dado of a room.

SURBASED ARCH.—An arch whose rise is less than the half-span.

SWEEP.—A thin piece of wood used in drawing curves. Any surface of easy curve.

T

T-SQUARE.—See Square.

TABLET.—A slab of stone, metal, or wood, bearing an inscription, and affixed to a building.

TABLING.—1. The coping of gable walls.—2. The indenting of the ends of the pieces forming a scarf, so that the joint will resist a longitudinal strain.

TACK.—A very small nail with flat head.

TÆNIA, Tenia.—The band or fillet which separates the Doric frieze from the architrave.

TAIL IN.—To fasten anything by one of its ends into a wall.

TAIL TRIMMER.—A trimmer next to the wall, into which the ends of joists are fastened to avoid flues.

TAMBOUR.—The naked part of Corinthian and Composite capitals, which bears some resemblance to a drum. Also, the wall of a circular temple surrounded with columns, and the circular vertical part of a cupola.—A cylindrical stone, such as one of the courses of the shaft of a column.

TANG.—The part of chisels and similar tools inserted in the handle.

TAPER SHELL-BIT. — A species of boring-bit used by joiners. It is conical both within and without, and its horizontal section is a crescent, the cutting edge being the meeting of the interior and exterior conical surfaces. Its use is for widening holes in wood.

TEAZE TENON.—A tenon on the top of a tenon, with a double shoulder and tenon from each, for supporting two level pieces of timber at right angles to each other.

TEMPLATE.—A short piece of timber, stone, &c., laid under the end of a beam or girder, resting on a wall, particularly in brick buildings, to distribute the weight over a large surface.

TEMPLET.—A pattern or mould used by masons, machinists, smiths, shipwrights, &c., for shaping anything by. It is made of tin or zinc plate, sheet-iron, or thin board, according to the use to which it is to be applied.

TENON.—The end of a piece of wood cut into the form of a rectangular prism, which is received into a corresponding cavity in another piece called a *mortise.—Tenon saw* (corruptly, *tenor saw*), a small saw with a brass or steel back, used for fine work, such as cutting tenons, dove-tails, mitres for joints, &c.

TENSION.—A pulling stress; thus, the rope or chain of a crane in operation is in tension.

TEREDO NAVALIS.—The *ship-worm,* which damages timber by boring into it.

TERMINUS.—A pillar statue; that is, a half statue, or bust, springing out of the square pillar which serves as its pedestal.

TETRAHEDRON.—A regular solid bounded by four equilateral triangles.

THRESHOLD.—A door-sill.

THROAT.—A channel or groove worked in the projecting part of the under side of a sill, coping, &c., to throw off the water and prevent it running inwards towards the wall.

TIE.—A timber string, chain, or rod of metal connecting and binding two bodies together which have a tendency to separate or diverge; such as *tie-beams, tie-rods,* &c.

Terminal Statue of Pan.   Antique Terminal Bust.

TILE.—A kind of thin brick or plate of baked clay, used for covering the roofs of buildings, and for paving floors, constructing drains, &c. Coloured tiles, glazed or enamelled, are used for floors, hearths, wall-linings, &c.

TILE-CREASING.—Two rows of plain tiles placed horizontally under the coping of a wall, and projecting about 1½ inch over each side, to throw off the rain-water.

TILTING-FILLET.—A fillet of wood, with one edge thicker than the other, laid under slating or tiling at the eaves or where it joins to a wall, to raise it slightly.

TIMBER.—1. That sort of wood which is squared, or capable of being squared, and fit for being employed in house or ship-building, or in carpentry, joinery, &c. The term is often restricted to pieces of wood of large scantling, in distinction from smaller stuff, such as planks, deals, battens, and boards. Unsquared timber is usually known as *round timber*. *Wood* is a general term, comprehending timber, dye-woods, fancy woods, fire-wood, &c., but the word *timber* is often used in a loose sense for all kinds of felled and seasoned wood.—2. A single piece or squared stick of wood for building, or already framed ; one of the main beams of a fabric.

TONGUE.—A small projection on the edge of a board made to fit into a corresponding groove in the edge of another board. A small strip of wood or iron made to fit into grooves in the edges of two contiguous boards.

TOOTHING.—Bricks or stones left projecting at the end of a wall, that they may be bonded into a continuation of it when required ; also, a tongue or series of tongues.

TOOTHING PLANE. — A plane, the iron of which is formed into a series of small teeth. It is used to roughen a surface intended to be covered with veneer or cloth, in order to give a better hold to the glue.

TOOTH-ORNAMENT. — An ornament characteristic of the Early English period of Gothic architecture. It is generally inserted in the hollow mouldings of doorways, windows, &c.

Tooth Ornament.

TORCH, v.—In plastering, to point the inside joints of slating laid on lath with lime and hair.

TORUS.—A large moulding used in the bases of columns. Its section is semicircular, and it differs from the astragal only in size, the astragal being much smaller.

TOTE.—The handle of a plane.

TRACERY.—That species of pattern work, formed or *traced* in the head of a window, by the mullions being there continued, but diverging into arches, curves, and flowing lines, enriched with foliations. Also, any ornamental design of the same character, for doors, panelling, or ceilings. See FAN-TRACERY.

TRACING CLOTH, TRACING PAPER.—Transparent cloth and paper used for copying drawings.

TRAMMEL.—An instrument for drawing ellipses.

TRANSEPT.—The transverse portion of a church built in the form of a cross ; that part which is placed between the nave and choir, and extends beyond their sides.

TRANSOM.—A horizontal bar of stone or timber across a mullioned window, dividing it into stories ; also, the cross-bar separating a door from the fan-light above it.

TRAPEZIUM.—A plane figure contained by four straight lines, none of them parallel.

TRAPEZOID.—A plane figure contained by four straight lines, two of them parallel.

TREAD.—The horizontal surface of a step.

TREFOIL. — An ornamental foliation much used in Gothic architecture in the tracery of windows, panels, &c. It consists of three cusps, the spaces enclosed between them producing a form similar to a three-lobed leaf.

Trefoils.

TRELLIS, TRELLIS-WORK. — A reticulated arrangement of wood laths, up which climbing plants, &c., can be trained. In *expanding trellis* the laths are merely pivoted at their intersections.

TRENAIL (commonly pronounced *trunnel*).—A cylindrical wooden pin.

TRIANGLE.—1. A plane figure with three sides and three angles.—2. A set-square.

TRIGLYPH.—An ornament used in the frieze of the Doric column, consisting of two entire **V**-shaped gutters or channels, called *glyphs*, and two half channels separated by three interspaces, called *femora*. See GUTTÆ.

TRIMMER-ARCH.—A brick arch for the support of a hearth in an upper floor. It is turned from the chimney breast to a joist parallel to it.—*Trimmer-joist, or Trimmer*, the joist against which the trimmer-arch abuts, or a transverse joist framed into a trimming-joist at one or both ends, and supporting the ends of intermediate joists.—*Trimming-joists*, the joists thicker than the common bridging-joists into which the trimmer is framed. Sometimes this term is applied only to trimming-joists which are parallel to the bridging-joists.

TRIMMING.—The working of any piece of timber into the proper shape, by means of the axe or adze. Also, the arrangement of joists, rafters, &c., around openings in floors, hearths, chimneys, sky-lights, &c.

TRINGLE.—In architecture, a little square member or ornament, as a listel, reglet, platband, and the like, but particularly a little member fixed exactly over every triglyph.

TROCHILUS.—The same as *scotia*.

TROUGH GUTTER.—A gutter in form of a trough ; an eaves-gutter.

TRUSS.—A combination of timbers, of iron, or of timbers and iron-work, so arranged as to constitute an unyielding frame. Such combinations are seen in roof-trusses, bridge-trusses, trussed beams, trussed partitions, &c.

TRYING-PLANE.—A plane used after the *jack-plane*, for taking off or shaving the whole length of the stuff, which operation is called *trying up*.

TUDOR FLOWER.—A trefoil ornament, common in Tudor architecture, and employed as a crest or ornamental finishing on cornices, ridges, &c.

Tudor Flower.

TUDOR STYLE.—Properly, the architecture which prevailed in England during the reigns of the Tudor family, from 1485 to 1603. The term is, however, generally restricted to the period which terminated with the death of Henry VIII., and may, perhaps, be most correctly designated as *late Perpendicular*. Its principal characteristics are the depressed, four-centred arch, and a peculiar dome-shaped turret. See ELIZABETHAN STYLE.

TUMBLED IN.—The same as *trimmed*.

TUMOUR.—A particular defect in trees.

TURNING.—The operation of cutting wood by means of chisels, &c., while the wood is revolved in a lathe.

TURNING PIECE.—A centre for a thin brick arch.

TUSCAN ORDER.—The simplest of the five Roman orders.

TUSK TENON. — A form of tenon used principally for securing floor-joists to trimmers, and trimmers to trimming-joists.

TWISTED.—Winding ; not straight.

TYMPAN, TYMPANUM.—The space in a pediment included between the cornice of the inclined sides and the fillet of the corona. Also the die of a pedestal, and the panel of a door.

U

ULCER.—A particular defect in trees.

UNDER-CROFT.—A vault under the choir of a church.

UNDERFOOT, UNDERPIN, v.—To support a wall, or a mass of earth or rock, when an excavation is made beneath it, by building up under it from the lower level.

V

VALLEY.—The re-entrant angle of a roof. — *Valley-gutter*, the channel of boards covered with lead formed in the valley of a roof.—*Valley-rafter*, the inclined timber supporting the jack-rafters which form a valley.

VAULT.—A continued arch, or an arched roof, so constructed that the stones, bricks, or other material of which it is composed sustain and keep each other in their places.

VAULTING SHAFT, VAULTING PILLAR.—A pillar sometimes rising from the floor to the spring of the vault of a roof ; more frequently, a short pillar attached to the wall, rising from a corbel, and from the top of which the ribs of the vault spring.

1. Cylindrical, barrel, or wagon vault. 2. Groined vault, formed by the intersection of two equal cylinders. 3. Gothic quadripartite vault, with groin-ribs. 4. Spherical or domical vault.

VENEER.—A facing of superior wood placed in thin leaves over an inferior sort. Generally, a facing of superior material laid over an inferior material.

VENETIAN WINDOW.—A window of large size divided by columns or pilasters into three lights, the middle one of which is usually wider than the others, and is sometimes arched.

VERANDA, VERANDAH. — A kind of open portico, or light external gallery in front of a building, with a sloping roof supported on slender pillars, and sometimes partly enclosed in front.

VERGE-BOARDS.—See BARGE-BOARDS.

VERSED SINE.—The height of an arc.

VESICA PISCIS.—A name given to a figure formed usually by the intersection of two equal circles, but often also assuming the form of an ellipse or an oval.

VESTIBULE.—1. The porch or entrance into a house, or a large open space before the door, but covered.—2. An ante-chamber before the entrance of an ordinary apartment; an apartment which serves as the means of communication to another room or series of rooms.

Vesica-piscis Seal, Wimborne Minster.

Vitruvian Scroll.

VITRUVIAN SCROLL.—A name given to a series of undulating scrolls joined together.

VOLUTE.—A kind of spiral scroll, used in the Ionic and Composite capitals, of which it is a principal ornament.

Volutes of the Ionic and Corinthian Capitals.

VOUSSOIR.—A stone in the shape of a truncated wedge which forms part of an arch. SEE ARCH.

W

WABBLE-SAW.—A circular saw hung out of truth on its arbor, used in cutting dove-tail slots, mortises, &c.

WAGON VAULTING. See CYLINDRICAL VAULTING.

WAINSCOT.—The timber-work that serves to line the walls of a room, and usually made in panels.— Wainscot Oak is oak of good quality sawn into boards in such a manner as to expose the silver grain.

WALES, or WALING-PIECES.—The horizontal timbers serving to connect a row of main piles together, and to support the boards in excavated trenches.

WALL-PLATE.—1. A piece of timber let into a wall to serve as a bearing for the ends of the joists.—2. A plate resting on a wall, and supporting the frame of the roof.

WALL-STRING.—The stair-string adjoining a wall.

WANE.—A rounded angle in squared timber, due to the attempt to obtain a larger scantling than the round log would allow.— Wany, containing wanes; not die-square.

WARPING. See CASTING.

WASH-BOARD.—The plinth or skirting of a room.

WASHER.—A flat piece of iron or other metal pierced with a hole for the passage of a bolt, between the nut of which and the timber it is placed to prevent undue pressure on the timber.

WEATHER, v.—To slope a surface, so that it may throw off the water.— Weather-boarding, boards laid horizontally on a vertical surface with a lap on each other, to prevent the penetration of rain and snow. These boards are generally made thinner on one edge than on the other, the thick edge of the upper board being laid to overlap the thin edge of that below. — Weather-moulding. 1. A moulded string-course.—2. The pro-

jecting moulding of an arch, having a weathered or sloped surface at top.

WEDGE.—A piece of wood or metal, thick at one end and sloping to a thin edge at the other, used for splitting wood, stone, &c., and for insertion in the joints of framing, &c. A tapering piece of wood (driven into a joint in brickwork, &c.) to which wood linings, skirtings, &c., can be nailed; a plug.

WELL-HOLE, WELL.—In a double flight of stairs, the space left in the middle between the ends of the steps.

WELSH-GROIN or UNDERPITCH GROIN.—A groin formed by the intersection of two cylindrical vaults, of which one is of less height than the other.

WET-ROT.—Decay in timber due to excessive moisture.

WHEEL-WINDOW.—See ROSE-WINDOW.

WHIP-SAW.—1. A thin narrow saw-blade strained in a frame and used in following curved lines.—2. A saw usually set in a frame for dividing or splitting wood in the direction of the fibres. It is wrought by two persons.

WHITE FIR or WHITE DEAL.—The produce of the Picea excelsa or Norway spruce.

WHITE PINE.—The produce of the North American Pinus Strobus, sold in this country as Yellow Pine.

WHITE WOOD.—Alburnum or sapwood.

WICKET.—A small door formed in a larger one, to admit of ingress and egress without opening the whole.

WILLESDEN PAPER.—A waterproof paper used instead of sarking felt, and for other purposes.

WIMBLE.—An instrument used by carpenters and joiners for boring holes; a kind of augur.

WIND.—To cast or warp; to turn or twist any surface, so that all its parts do not lie in the same plane.— Winding, a surface whose parts are twisted so as not to lie on the same plane. When a surface is perfectly plane it is said to be out of winding.— Winding-sticks, two slips of wood, each straightened on one edge and having the opposite edge parallel. Their use is to ascertain whether the surface of a board, &c., winds or is twisted.

WIND-BEAM.—1. An old name for collar-beam.—2. A piece of wood laid diagonally under the rafters of a long roof from the foot of one truss to the head of another, to strut them and so prevent the roof from being racked with the wind.

WINDERS.—Those steps of a stair which, radiating from a centre, are narrower at one end than at the other.— Winding-Stairs, stairs ascending spirally round a solid or an open newel.

WINDOW.—An opening in the wall of a building and fitted with a glazed frame of wood or iron, for the admission of light, and of air when necessary.— Window-board, the horizontal board covering the sill of stone, brick, or other material inside a room. Sometimes known as window-bottom or window-sill.— Window-frame, the frame of a window which receives and holds the sashes or casements.— Window-linings, the boards or panelling used for covering the internal reveals and soffits of window-openings.— Window-sash, the sash or light frame in which panes of glass are set for windows. See SASH.

WING.—A smaller part attached to the side of the main edifice.

WING-COMPASS.—A joiner's compass with an arc-shaped piece that passes through the opposite leg, and is clamped by a screw.

WOOD-BRICKS.—Blocks of wood of the shape and size of bricks, inserted in the interior walls of a building as holds for the joinery. Also, blocks of a somewhat similar shape and size used for the paving of streets.

WOOD-WORKING MACHINERY.—Machines for operating on wood by sawing, planing, moulding, &c., in distinction from hand-tools.

WREATHED.—A term applied to stair-strings, handrails, &c., worked on the curve.

Y Z

YELLOW DEAL.—See DEAL.

YELLOW PINE.—The American name for the Southern Pines, sold in this country as Pitch-pine. See WHITE PINE.

ZIGZAG MOULDING. See CHEVRON.

ZOCCO, ZOCLE, ZOCOLO. See SOCLE.

INDEX

The Modern Carpenter Joiner and Cabinet-Maker series is divided into two major volumes, each comprising four divisional volumes. Each of the two major volumes has been indexed separately. For the purposes of indexing the series, i refers to major volume one, comprising divisional volumes I, II, III, and IV, while ii refers to major volume two, comprising divisional volumes V, VI VII, and VIII.